Vaccine Development

Edited by Yulia Desheva

Published in London, United Kingdom

IntechOpen

Meet the editor

Dr. Yulia Desheva is a leading researcher at the Institute of Experimental Medicine, St. Petersburg, Russia. She is a professor in the Stomatology Faculty, St. Petersburg State University. She has expertise in the development and evaluation of a wide range of live mucosal vaccines against influenza and bacterial complications. Her research interests include immunity against influenza and COVID-19 and the development of immunization schemes for high-risk individuals.

Contents

Preface

Infectious diseases are some of the most common causes of morbidity and mortality. Vaccination is the most effective and scientifically based means of protection against preventable infections. This has been proven by more than 200 years of experience in vaccination, starting from the time of Edward Jenner (1749–1823), who developed the smallpox vaccine, which was the world's first ever vaccine. The use of vaccines has made it possible to defeat several dangerous viral infections such as polio and rabies.

This book includes a series of chapters that highlight some issues of human and animal vaccine development against several bacterial, viral, and parasitic infections.

Currently, a global problem is the widespread coronavirus infection caused by SARS-CoV-2, which is actively mutating. The usual path of vaccine development from antigen selection through preclinical studies and clinical trials to production can take more than ten years. To develop a vaccine against COVID-19 as quickly as possible, scientists and developers have worked in parallel at several different stages. This confirms that vaccine research is always relevant and must continue.

The authors hope that this book will be interesting for a wide range of readers interested in the theory and practice of developing vaccines and vaccine prevention.

I would like to thank all the authors for their excellent contributions. I dedicate this book to my parents who provided invaluable support during the process of preparing this volume for publication. I am also grateful to the staff at IntechOpen for their prompt assistance at all stages of this work.

Yulia Desheva
Head of Department of Translation Medicine,
FSBSI 'Institute of Experimental Medicine',
Saint Petersburg, Russian Federation

Professor,
Saint Petersburg State University,
Saint Petersburg, Russian Federation

Section 1

Introduction

Chapter 1

Introductory Chapter: Current Trends in Vaccine Development

Yulia Desheva

1. Introduction

Ever since Edward Jenner proposed in 1796 a prophylactic inoculation of cowpox using the lymph of a diseased animal as a vaccine [1], vaccination of humans has become widespread. Currently, millions of people around the world receive vaccines against viral and bacterial infections. The first generation of vaccines were whole particles of pathogens, attenuated and inactivated in various ways. The whole virion lives attenuated and inactivated vaccines have the longest history of use. Inactivated vaccines contain non-viable viruses, and repeated injections are often required to form an immune response. Although, the experience of eliminating dangerous viral infections, such as smallpox, measles, and poliomyelitis suggests that the use of live vaccines provides the necessary epidemic efficiency and effectiveness of anti-epidemic measures.

Live attenuated influenza vaccines (LAIVs) have been used in Russia since the 1960s. In the early days of LAIV production, the serial passage in chick embryos (CE) produced attenuated viruses, which were host-range (hr-) mutants. Serial passages in CE at a temperature lowered to 25–28°C made it possible to regularly obtain cold-adapted (ca-) vaccine strains from all varieties of influenza A and B viruses which were safe for children [2]. Based on data on the segmented genome of the influenza virus, it became possible to develop vaccine strains production using genetic reassortment of epidemic influenza viruses and an attenuated master donor strain (MDS) [3]. The main features of reassortant vaccine strains—safety for susceptible people and genetic stability—are determined by the properties of MDS. A cold-adapted MDS A/Leningrad/134/57 (H2N2) is currently used in Russia to prepare the reassortant A/H1N1 and A/H3N2 LAIV vaccine strains [4]. The MDS B/USSR/60/69 was developed for the production of vaccine strains of type B influenza viruses [5]. In 2003, the American LAIV was manufactured by MedImmune. Inc. was licensed for use in North America and Europe [6]. LAIV has been shown to be particularly effective in preventing influenza among young children [7]. Differences between Russian and American MDSs are both in the total number of passages and in the cell model.

Until recently, influenza viruses were considered the main causative agents of pandemics and annual epidemics in the modern world. Influenza epidemics have resulted in approximately 3–5 million cases of severe infection and between 290,000 and 650,000 deaths worldwide annually [8]. New coronavirus infection disease (COVID-19), which was first reported on December 31, 2019, in Wuhan, China, caused more than 378 million cases and more than 5.6 million deaths worldwide [https://github.com/CSSEGISandData/COVID-19] as have been reported on February 1st, 2022.

Back in 2015, a chimeric virus expressing the spike of bat coronavirus SHC014 based on a mouse-adapted severe acute respiratory syndrome coronavirus

(SARS-CoV) backbone was reported to have been able efficiently bind to human ACE2 receptor and replicate efficiently in primary human airway cells. In mice, the double-inactivated whole SARS-CoV vaccine failed to neutralize and protect from infection with CoVs with the novel spike protein [9]. However, the use of modern technologies has made it possible to prepare and apply a number of vaccines against this infection in the shortest possible time.

A large part of the development of prototypes of COVID-19 vaccines is based on the use of viral vectors. The technology for this approach is based on the insertion of a gene encoding the target viral protein into the genome of a viral vector. A vector is another virus that does not cause disease in humans. For example, to create a vaccine against COVID-19, a gene encoding a coronavirus protein is inserted into an adeno-virus [10]. An obstacle to the use of such vaccines is the presence of antibodies to the viral vector in humans. In this case, to obtain a full-fledged immune response is to adjust vaccination regimens, such as priming boosting is applied.

The production of vaccines based on both non-replicating mRNA and self-amplifying mRNA is a promising new direction in vaccinology. DNA and RNA vaccines are preparations based on the nucleic acid to deliver viral genetic material into the cells of the body. This was limited the use of DNA and RNA vaccine as and for a long time no nucleic acid-based preparation has been used in clinical practice in humans. The emergence of the COVID-19 pandemic has dramatically accelerated this research and RNA vaccines are now widely used.

There are several advantages of mRNA vaccines over conventional methods, namely: minimization of the potential risk of infection and mutagenesis due to the natural degradation of mRNA in the cellular microenvironment; high efficiency of the immunogen due to structural modifications of mRNA increases its stability and translation efficiency; highly effective mRNA-based vaccines able to generate anti-viral neutralizing antibodies; recombinant mRNA facilitates large-scale production of sufficient doses vaccines needed to treat large population groups. These factors make an mRNA vaccine more suitable for rapid pandemic response [11].

The SARS-Cov-2 mutates rapidly, with the formation of new variants with increased transmissibility such as Delta (B.1.617.2 and AY lineages) and Omicron (B.1.1.529 and BA lineage [https://www.cdc.gov/coronavirus/2019-ncov/variants/variant-classifications.html#anchor_1632158924994]. Therefore, the development of universal or polyepitope vaccines is relevant. Antigen selection is a key aspect of any vaccine design. Developing computer and bioinformatic technologies and applied mathematical analysis help to predict antigenic determinants SARS-CoV-2 and immune responses to them. The concept of a multi-epitope vaccine against SARS-CoV-2 is in the identification and assembly of epitopes for B cells, CTL CD8+ and T-helpers CD4 + ,B- and T-cell epitopes into a single immunogen, which can induce a more effective response of both parts of immunity—humoral and cellular [12].

Another concern with the high variability of SARS-Cov-2 is the possible need for frequent and repeated vaccinations. This raises the question of which vaccines will be safe and effective for these purposes. In general, the information accumulated to date on the pathogenesis of COVID-19 infection indicates the central role of the mucous membranes and mucosa-associated lymphoid tissues in the initiation, clinical development, and spread of the [13]. As evidenced by studies of taxonomically and structurally similar coronaviruses (SARS-CoV and MERS-CoV), mucosal vaccination may provide a safe and effective way to induce not only long-term mucosal immunity but also systemic immune protection against SARS-CoV-2 [14, 15]. It is considered that genetically modified microorganisms, including probiotic strains, are attractive agents for oral administration or mucosal delivery of vaccine antigens. Several studies have shown that mucosal administration of antigens is capable of eliciting an immune response mediated by mucosal-specific serum IgG and IgA along with mucosal cell-mediated immune responses that effectively neutralize and eradicate infections [16–18]. Thus, advances in modulating mucosal immune responses, and in particular the use of probiotics as live delivery vectors, may encourage prospective studies to evaluate the efficacy of genetically engineered probiotics in SARS-CoV-2 infection [19, 20]. Controversial aspects of the use of genetically modified probiotics lie in overcoming interference between mucosal delivery of therapeutic agents and the immune system.

Along with this, it is worth mentioning the development of vaccines against bacterial infections. The number of bacterial vaccines against common pathogens such as whooping cough, *Streptococcus pneumoniae,* and *Haemophilus influenzae* are also widely used and updated. Pneumococcal polysaccharide or conjugate vaccines are currently used, which are based on a limited set of polysaccharide antigens of the bacterial capsule. Although, polysaccharide vaccines require a constant change in composition in view of the fact that when vaccinating the population, bacterial serotypes that are not included in the vaccine begin to dominate and cause disease [21, 22]. Therefore, pneumococcal vaccines include more and more components, for example, pneumococcal conjugate vaccine Prevnar was first prepared as a 7-component preparation, then as a 13-component preparation, and now it is being prepared as a 20-component preparation [23]. The inclusion of a large number of components and their modification with a protein adjuvant makes the vaccine difficult to control in terms of quality and safety, and also expensive. In addition, the polysaccharide components of the vaccine do not provide a long immunological memory, which requires their modification with adjuvants or repetitive revaccination [23]. Current trends in the development of vaccines against pneumococcal infection include the use of protein factors of bacterial pathogenicity, such as highly conservative lipoprotein or enzymes (nucleases, proteases, hemolysin, peptidase) which are preferable over polysaccharide vaccines due to the lower variability and higher immunogenicity [24, 25].

Finally, a number of areas include the development of vaccines against a number of somatic diseases, which, as it turned out, are closely related to a number of pathogens. For example, even 50 years ago, no one could have imagined that such diseases as stomach and duodenal ulcers are of infectious origin and are related to *Helicobacter pylori*. The development of an anti-*Helicobacter pylori* vaccine turned out to be a rather complicated project due to the number of pathophysiological, immunological, and technological difficulties. Nevertheless, a promising direction in improving *H. pylori* vaccines is the search for effective mucosal adjuvants and the use of immunostimulatory probiotics during the administration of a vaccine. Perhaps in the future, it will be possible to prevent somatic diseases by vaccinating against the corresponding infections.

In conclusion, it should be noted that vaccination is the most effective and cost-effective means of protection against infectious diseases known to modern medicine. The use of vaccines has reduced, and in some cases completely eliminated, a number of diseases from which tens of thousands of children and adults previously suffered and died. Vaccines represent the most powerful defense tool in reducing the risk of a pandemic outbreak and will play a critical role in response to any future pandemic. The development and testing of new vaccine platforms contribute to the rapid release of vaccines against new pathogens that continually arise and regenerate. To fight new emerging infections, the creation and deposit of prototype vaccine candidates can also help. When developing new vaccines, the future of the drug is determined by such factors as the possibility of reducing the incidence and the benefits of using the vaccine, the risk of complications and possible damage from vaccination, the cost of the vaccine, and economic benefits. The cost of vaccination for any vaccine with proven efficacy is about 10 times less than the cost of treating infectious diseases.

Author details

Yulia Desheva[1,2]

1 Federal State Budgetary Scientific Institution "Institute of Experimental Medicine", Saint Petersburg, Russian Federation

2 Department of Fundamental Problems of Medicine and Medical Technologies, Saint Petersburg State University, Saint Petersburg, Russian Federation

*Address all correspondence to: desheva@mail.ru

IntechOpen

References

[1] Riedel S. Edward Jenner and the history of smallpox and vaccination. Proceedings (Baylor University. Medical Center). 2005;**18**(1):21-25. DOI: 10.1080/08998280.2005.11928028

[2] Smorodintsev AA, Dokuchaev GI, Minichev PN, Filippov NA, Chalkina OM. Epidemiologic efficacy of live influenza vaccine during 1962 outbreaks of influenza A2 and B. Federation Proceedings. Translation Supplement; Selected Translations from Medical-related Science. 1966;**25**(5):829-832

[3] Alexandrova GI, Budilovsky GN, Koval TA, Polezhaev FI, Garmashova LM, Ghendon YZ, et al. Study of live recombinant cold-adapted influenza bivalent vaccine of type A for use in children: An epidemiological control trial. Vaccine. 1986;**4**(2):114-118. DOI: 10.1016/0264-410X(86)90049-6

[4] Kendal AP, Maassab HF, Alexandrova GI, Ghendon YZ. Development of cold-adapted recombinant live, attenuated influenza A vaccines in the USA and USSR. Antiviral Research. 1982;**1**(6):339-365. DOI: 10.1016/0166-3542(82)90034-1

[5] Alexandrova GI, Maassab HF, Kendal AP, Medvedeva TE, Egorov AY, Klimov AI, et al. Laboratory properties of cold-adapted influenza B live vaccine strains developed in the US and USSR, and their B/Ann Arbor/1/86 cold-adapted reassortant vaccine candidates. Vaccine. 1990;**8**(1):61-64. DOI: 10.1016/0264-410x(90)90179-p

[6] Pebody R, McMenamin J, Nohynek H. Live attenuated influenza vaccine (LAIV): Recent effectiveness results from the USA and implications for LAIV programmes elsewhere. Archives of Disease in Childhood. 2018;**103**(1):101-105

[7] Chung JR, Flannery B, Ambrose CS, Bégué RE, Caspard H, DeMarcus L, et al. Live attenuated and inactivated influenza vaccine effectiveness. Pediatrics. 2019;**1**:143(2)

[8] Iuliano AD, Roguski KM, Chang HH, Muscatello DJ, Palekar R, Tempia S, et al. Estimates of global seasonal influenza-associated respiratory mortality: A modelling study. The Lancet. 2018;**391**(10127):1285-1300

[9] Menachery VD, Yount BL, Debbink K, Agnihothram S, Gralinski LE, Plante JA, et al. A SARS-like cluster of circulating bat coronaviruses shows potential for human emergence. Nature Medicine. 2015 Dec;**21**(12):1508-1513

[10] Mendonça SA, Lorincz R, Boucher P, Curiel DT. Adenoviral vector vaccine platforms in the SARS-CoV-2 pandemic of the impact and development of this emerging platform. NPJ Vaccines. 2021;**6**(1):1-4. DOI: 10.1038/s41541-021-00356-x

[11] Wang F, Kream RM, Stefano GB. An evidence-based perspective on mRNA-SARS-CoV-2 vaccine development. Medical Science Monitor. 2020;**26**:e924700. DOI: 10.12659/MSM.924700

[12] Enayatkhani M, Hasaniazad M, Faezi S, Gouklani H, Davoodian P, Ahmadi N, et al. Reverse vaccinology approach to design a novel multi-epitope vaccine candidate against COVID-19: An in silico study. Journal of Biomolecular Structure and Dynamics. 2021;**39**(8):2857-2872

[13] Gallo O, Locatello LG, Mazzoni A, Novelli L, Annunziato F. The central role of the nasal microenvironment in the transmission, modulation, and clinical progression of SARS-CoV-2 infection. Mucosal Immunology. 2020;**14**(2):305-316. DOI: 10.1038/s41385-020-00359-2 34

[14] Moreno-Fierros L, García-Silva I, Rosales-Mendoza S. Development of SARS-CoV-2 vaccines: Should we focus on mucosal immunity? Expert Opinion on Biological Therapy. 2020;**20**(8):831-836. DOI: 10.1080/14712598.2020. 1767062 35

[15] Mudgal R, Nehul S, Tomar S. Prospects for mucosal vaccine: Shutting the door on SARS-CoV-2. Human Vaccines & Immunotherapeutics. 2020;**16**(12):2921-2931. DOI: 10.1080/21645515.2020.1805992

[16] Bermúdez-Humarán LG, Kharrat P, Chatel J-M, Langella P. Lactococci and lactobacilli as mucosal delivery vectors for therapeutic proteins and DNA vaccines. Microbial Cell Factories. 2011;**10**:S4. DOI: 10.1186/1475-2859-10-S1-S4

[17] De Azevedo M, Karczewski J, Lefévre F, Azevedo V, Miyoshi A, Wells JM, et al. In vitro and in vivo characterization of DNA delivery using recombinant *Lactococcus lactis* expressing a mutated form of *L. monocytogenes* Internalin A. BMC Microbiology. 2012;**12**:299. DOI: 10.1186/1471-2180-12-299

[18] Gupalova T, Leontieva G, Kramskaya T, Grabovskaya K, Kuleshevich E, Suvorov A. Development of experimental pneumococcal vaccine for mucosal immunization. PLoS One. 2019;**14**(6):e0218679. DOI: 10.1371/journal.pone.0218679

[19] Taghinezhad-S S, Mohseni AH, Bermúdez-Humarán LG, Casolaro V, Cortes-Perez NG, Keyvani H, et al. Probiotic-based vaccines may provide effective protection against COVID-19 acute respiratory disease. Vaccine. 2021;**9**(5):466

[20] Suvorov A, Gupalova T, Desheva Y, Kramskaya T, Bormotova E, Koroleva I, et al. Construction of the novel vaccine candidate against SARS-Cov-2 based on enterococcal probiotic. Frontiers in Pharmacology;**2022**:3753. DOI: 10.3389/fphar.2021.807256

[21] Jose RJ, Periselneris JN, Brown JS. Community-acquired pneumonia. Current Opinion in Pulmonary Medicine. 2015;**21**(3):212-218

[22] Chalmers JD, Campling J, Dicker A, Woodhead M, Madhava H. A systematic review of the burden of vaccine preventable pneumococcal disease in UK adults. BMC Pulmonary Medicine. 2016;**16**(1):77

[23] US Food and Drug Administration. US FDA Approves PREVNAR 20™, Pfizer's Pneumococcal 20-valent Conjugate Vaccine for Adults Ages 18 Years or Older. Silver Spring, Maryland, United States: US Food and Drug Administration; 2021

[24] Dorosti H, Eslami M, Negahdaripour M, Ghoshoon MB, Gholami A, Heidari R, et al. Vaccinomics approach for developing multi-epitope peptide pneumococcal vaccine. Journal of Biomolecular Structure and Dynamics. 2019;**37**(13): 3524-3535

[25] Suvorov A, Dukhovlinov I, Leontieva G, Kramskaya T, Koroleva I, Grabovskaya K, et al. Chimeric protein PSPF, a potential vaccine for prevention Streptococcus. Vaccines and Vaccination. 2015;**6**(6):304. DOI: 10.4172/2157-7560.1000304

Section 2

Modern Trends in the Design and Delivery of Vaccines

Chapter 2

In Silico Vaccine Design Tools

Shilpa Shiragannavar and Shivakumar Madagi

Abstract

Vaccines are a boon that saves millions of lives every year. They train our immune system to fight infectious pathogens. According to the World Health Organization, vaccines save 2.5 million people every year and protect them from illness by decreasing the rate of infections. Computational approach in drug discovery helps in identifying safe and novel vaccines. *In silico* analysis saves time, cost, and labor for developing the vaccine and drugs. Today's computational tools are so accurate and robust that many have entered clinical trials directly. The chapter gives insights into the various tools and databases available for computational designing of novel vaccines.

Keywords: tools, databases, computational approach, reverse vaccinology

1. Introduction

The vast genome information obtained after the sequencing projects has paved the way to several *in silico* screening and computational analysis [1]. Nowadays, the vaccine design approaches are based on computational analysis as the methods are time and cost-effective [2].

Bioinformatics is a field that uses information technology and mathematical elements to manage, analyze, and use biological data It is a field that is constantly evolving and producing useful tools for biological sciences [3]. The development and implementation of computational algorithms and software tools aid in the understanding of biological processes, with the primary goal of serving the agriculture and pharmaceutical industries, health care, forensic analysis, crop improvement, food analysis, drug discovery, and biodiversity management [4].

There are various *in silico* methods for studying linear B-cell epitopes, helper T lymphocytes (HTL), and cytotoxic T lymphocytes (CTL) epitopes [5]. Antigenicity, human population coverage, physicochemical properties, toxicity, allergenicity, and secondary structure of the designed vaccine are all evaluated using cutting edge bioinformatic approaches, to ensure that the designed vaccine is of high quality [6]. *In silico* tools are also available for prediction, refinement, and validation of the three-dimensional (3D) structures of the designed vaccine candidates [7].

2. *In silico* vaccine tools

VaxiJen is the first server to predict protective antigens without using alignment. It was created to allow antigen classification based solely on the physicochemical properties of proteins, rather than sequence alignment [8]. Vaxijen is a new alignment-free antigen prediction method based on auto-cross covariance (ACC) transformation of protein sequences into uniform vectors of major amino acid properties. Datasets were generated for viral, bacterial, and tumor proteins. There are validation models to evaluate the results [9].

The server can be used alone or in tandem with alignment-based prediction methods. It is freely available online at the following address: http://www.jenner.ac.uk/vaxijen.

VaxiJen is currently the only tool that can classify protein sequences exclusively based on their physicochemical properties, without any functional or biological information. It is a very fast and easy-to-use tool. The user's unable to alter the training dataset from which the prediction model is created remains the disadvantage of the tool [10].

2.1 ANTIGENpro

It uses two-stage architecture, multiple representations of the primary sequence, and five machine learning algorithms to predict the antigenicity of proteins [11]. ANTIGENpro is a sequence-based, alignment-free, and pathogen-independent predictor. An Support Vector Machine (SVM) classifier summarizes the results of the analysis and predicts if a protein is antigenic or not, as well as the corresponding probability [12]. The protein antigenicity predictor ANTIGENpro is the first to use reactivity data obtained *via* protein microarray analysis for five pathogens to determine whole protein antigenicity [11].

2.2 AllergenFP

The AllergenFP is developed based on the dataset that is described by five E-descriptors and the strings that were turned into uniform vectors through auto-cross covariance transformation [12].

Based on the statical analysis of sensitivity and specificity, the overall accuracy confirms that AllergenFP and AllerTOP are the best allergen prediction tools for sequencing compared to the other analysis tools and servers [13].

2.3 AllerTOP

A protein sequence is transformed by auto cross covariance (ACC) into uniform equal length vectors, a protein sequence mining technique developed by Wold *et al.*, 1993.

This technique was used to study quantitative structure–activity relationships (QSARs) of peptides of different lengths. The main characteristics of amino acids were expressed as five E descriptors. The data reveal the hydrophobicity of amino acids, molecular size, helix-forming tendency, the relative abundance of amino acids, and ability to form β-strands. A k-nearest neighbor algorithm (kNN, k = 1) is used to classify proteins based on a training set containing 2427 allergens from different species and 2427 non-allergens [14].

AllerTOP v.2 is a handy, robust, and highly complementary allergen prediction tool with the highest level of precision.

2.4 T- and B-cell epitope identification

The essential for epitope-based antibodies is the accessibility of epitopes. The idea of epitopes present in an antigen should be perceived for such immunization plan. There is a contrast between the acknowledgment of epitopes by B and T cells

[15]. B-cell receptors can tie to epitopes in antigen present either in dissolvable structure or on the outside of microbe, and there is no necessity of intercession by some other particle for this limiting [16]. Notwithstanding, the limiting instrument for T-cell epitopes is extraordinary, as they require an epitope to be introduced by MHC particles for restricting to the T-cell receptor [17].

B-cell epitopes are situated on the local protein and are both consistent and conformational. The consistent epitopes are otherwise called straight, or consecutive epitopes involve amino acids present successively in the protein [18]. B-cell epitopes are for the most part surface available, hydrophilic, polar locales of the antigens that can promptly tie to the individual counteracting agent particle [19].

Dissimilar to B-cell epitopes that can be perceived directly, T-cell epitopes require a show of epitope with MHC atoms. Lymphocyte epitopes are just direct or consecutive and the antigens need to go through handling prior to being perceived by their receptors [20]. The protein is the first to cut into little peptides; these peptides tie to MHC particles and hence structure a trimolecular complex with T-cell receptors. There are two kinds Tc cells or cytotoxic T cells that show CD8 protein particles on their surface and Th cells and T-helper cells showing CD4 surface protein. The epitopes that are introduced to Tc cells are shown by Class I MHC particles, while Th-cell epitopes are shown by Class II MHC atoms. The pathways of preparing and introducing epitopes to the two sorts of T cells are unique [21].

2.5 IEDB

The Immune Epitope Database and Analysis Resource (IEDB) is an unreservedly accessible asset that contains a broad assortment of tentatively estimated invulnerable epitopes and a setup of apparatuses for anticipating and dissecting epitopes [22]. The IEDB incorporates counteracting agent and T-cell epitopes for irresistible sicknesses, allergens, and immune system illnesses, and relocate to alloantigen concentrated in people, nonhuman primates, mice, and other species. Life science specialists can utilize the IEDB to foster new antibodies, diagnostics, and therapeutics. The dataset is populated utilizing data caught or curated from peer-reviewed and from information put together by scientists. As of December 2016, more than 18,000 references have been curated, and the dataset contains more than 260,000 epitopes and more than 1,200,000 B cell, T cell, MHC restricting, and MHC ligand elution tests [23].

A comprehensive list of freely available tools for determining the binding affinity of peptides in a protein to different MHC molecules is given in the following **Table 1.** These tools use machine learning methods such as hidden Markov models (HMMs), support vector machines (SVMs), position-specific scoring matrices (PSSMs), and artificial neural networks (ANNs).

2.6 BCPred

A continuous B-cell epitope prediction method uses support vector machine (SVM) classifiers that were trained on a homology-reduced dataset of 701 linear B-cell epitopes recovered from the Bcipep database and 701 non-epitopes randomly retrieved from Swiss-Prot sequences using five different kernel approaches and fivefold cross-validation [41].

The advantages of the tool include the samples taken in both the training and test datasets that were experimentally determined. Deep learning methods were implemented for making predictions and allowing the large number of datasets that can improve statistical analysis and features of B-cell epitopes [42].

Name of the tool	Description of tool
NetMHC for MHC-I	Prediction of peptide–MHC class I binding using artificial neural networks (ANNs). ANNs have been trained to recognize 81 different Human MHC alleles, including HLA-A, B, C, and E. Predictions for 41 animal alleles (Monkey, Cattle, Pig, and Mouse) are also available [24].
NetMHCPan for MHC-I	NetMHCpan 4.1 is the latest version of NetMHCpan. Using artificial neural networks, the NetMHCpan-4.1 server predicts peptide binding to any MHC molecule of known sequence based on ANNs. The method is trained on over 850,000 quantitative Binding Affinity (BA) and Mass-Spectrometry Eluted Ligands (EL) peptides. BA data includes 201 MHC molecules from humans (HLA-A, B, C, and E), mice (H-2), cattle (BoLA), and primates (P). By uploading a full-length MHC protein sequence, the user can obtain predictions for any custom MHC class I molecule. For peptides of any length, predictions can be made [25].
SYFPEITHI for both MHC I & II	It is constantly updated and comprises a library of MHC class I and class II ligands and peptide motifs from humans and other animals, such as apes, cattle, chicken, and mice. Individual entries are available for all of the motifs that are currently available. MHC alleles, MHC motifs, natural ligands, T-cell epitopes, source proteins/organisms, and references can all be found using this method [26].
ProPred for MHC-II	ProPred is a web-based graphical tool that predicts MHC class II binding regions in antigenic protein sequences. The server uses an amino-acid position coefficient database derived from literature to create a matrix-based prediction algorithm. Predicted binders can be viewed as peaks in the graphical interface or as colored residues in the HTML interface [27].
RANKPEP for MHC-I & II	RANKPEP is an online resource that predicts peptide–MHC class I binding using position specific scoring matrices (PSSMs) or profiles as a basis for CD8 T-cell epitope identification. RANKPEP has been extended to predict peptide-MHCII binding and anticipate CD4 T-cell egress using PSSMs that are structurally consistent with the binding mode of MHC class II ligands [28].
EpiJen for MHC-I	EpiJen is a reliable and consistent multi-step algorithm for T-cell epitope prediction that belongs to the next generation in silico T-cell epitope identification methods. These methods aim to reduce subsequent experimental work by increasing the success rate of epitope prediction [29].
MHCPred for MHC-I & II	A quantitative T-cell epitope prediction server. MHCPred includs alleles from the human leukocyte antigen A (HLA-A) locus. The server currently contains 11 human HLA class I, three human HLA class II, and three mouse class I models. In addition, the new MHCPred includes a binding model for the human transporter associated with antigen processing (TAP). The server also includes a tool for designing heteroclitic peptides. A confidence p value is used to improve the predictability of binding affinities [30].
MULTIPRED2 for MHC-I & II	MULTIPRED2 is a computer programme that predicts peptide binding to numerous alleles of the human leukocyte antigen class I and class II DR molecules with ease. Peptide binding to products of individual HLA alleles, combinations of alleles, or HLA supertypes can be predicted. As prediction engines, NetMHCpan and NetMHCIIpan are used. A1, A2, A3, A24, B7, B8, B27, B44, B58, B62, C1, and C4 are the 13 HLA Class I supertypes. DR1, DR3, DR4, DR6, DR7, DR8, DR9, DR11, DR12, DR13, DR14, DR15, and DR16 are the 13 HLA Class II DR supertypes. MULTIPRED2 predicts peptide binding to 1077 variations representing 26 HLA supertypes in total. It currently calculates population coverage in North America's five major groups. For the identification of T-cell epitopes, MULTIPRED2 is a useful tool to complement wet-lab experimental approaches [31].
NetMHCII for MHC-II	NetMHCII is an allele-specific approach that uses information from all MHC molecules in the data set. Using artificial neuron networks, the NetMHCII predicts peptide binding to HLA-DR, HLA-DQ, HLA-DP, and mouse MHC class II alleles. For 25 HLA-DR alleles, 20 HLA-DQ, 9 HLA-DP, and 7 mouse H2 class II alleles, predictions can be made. The prediction values are given as a percent -Rank to a set of 1,000,000 random natural peptides in nM IC50 values. The presence of strong and weak binding peptides is indicated [32].

Name of the tool	Description of tool
NetMHCIIPan for MHC-II	NetMHCIIpan is a pan-specific method that uses information from all MHC molecules in the data set. Using Artificial Neural Networks (ANN), the NetMHCIIpan-4.0 server predicts peptide binding to any MHC II molecule with a specified sequence (ANNs). It was trained on a large dataset of over 500,000 Binding Affinity (BA) and Eluted Ligand mass spectrometry (EL) measurements, which included the three human MHC class II isotypes HLA-DR, HLA-DQ, and HLA-DP, as well as mouse molecules (H-2). Peptides of any length can be predicted by the network. The model generates a prediction score for the chance of a peptide being delivered naturally by an MHC II receptor of choice. Percent rank score is also included in the output, which normalizes prediction score by comparing it to the prediction of a group of random peptides [33].
MHC2Pred for MHC-II	Promiscuous MHC class II binding peptides are predicted using an SVM-based technique. For 42 alleles, the average accuracy of the SVM-based technique is 80%. Because the dataset was smaller, the method's performance was lower for a few alleles. The method's performance was evaluated using 5-fold cross-validation [34].
ClustiMer	Instead of being distributed randomly throughout protein sequences, potential T-cell epitopes typically aggregate in specific immunogenic consensus sequence (ICS) regions as clusters of 9–25 amino acids with 4–40 binding motifs. The ClustiMer algorithm, in conjunction with EpiMatrix, can be used to identify peptides with EpiMatrix immunogenicity cluster scores of +10. These peptides are typically immunogenic [35].
NERVE	Predicts the best vaccine candidates based on a prokaryotic pathogen's flat file proteome. It is a fully automated reverse vaccinology system designed to predict best vaccine candidates from bacteria proteomes as well as manage and display data through user-friendly output [36].
BlastiMer	One can also use the BlastiMer programme to automatically BLAST "potential epitopes against the human sequence database at GenBank". BLASTing excludes epitopes with potential autoimmunity and cross-reactivity questions and locates the epitopes that can be used safely in the development of human or animal vaccines BlastiMer can also perform BLAST searches against the PDB, SwissProt, PIR, PRF, and non-redundant GenBank CDS translations [37, 38].
EpiMatrix	EpiVax, an in-silico tool designed to predict and identify the immunogenicity of therapeutic proteins and epitopes. It is also used to redesign proteins and create T-cell vaccines [39].
IEDB Population Coverage analysis	It determines the percentage of people in that location who have shown possible responses to the query epitopes. The Population Coverage Calculation programme is simple and flexible to use. Using MHC binding or T cell restriction data and HLA gene frequencies, a method for calculating anticipated population coverage of a T-cell epitope-based diagnostics are implemented [40].

Table 1.
List of various insilico vaccine design tools.

2.7 ABCpred

The ABCpred server uses an artificial neural network to anticipate linear B-cell epitope areas in an antigen sequence. This server will aid in the identification of epitope regions that can be used to choose synthetic vaccine candidates, diagnose diseases, and conduct allergy research. It is the first server developed based on a recurrent neural network with the fixed length patterns. The training and test-ing dataset has 700 B-cell and 700 non-B-cell epitopes or random peptides with a maximum length of 20 amino acids. About 65.93% accuracy was achieved using a recurrent neural network [43].

2.8 BepiPred

BepiPred is based on a random forest algorithm that is trained using epitopes from antibody–antigen protein structures. It is a new method based on known 3D structures and the large number of linear epitopes available from the IEDB database hence remains to outperform compared to the other tools. It presents results in a style that is both user friendly and useful for both computer experts and non-experts [44].

2.9 LBtope

It is developed based on the experimentally validated B-cell epitopes and non-B cell epitopes from IEDB. Two types of datasets were derived as LBtope variable with 14,876 and 23,321 B-cell epitopes and non-epitopes of variable lengths, whereas LBtope fixed length has datasets with 12,063 B-cell epitopes and 20,589 non-epitopes of fixed lengths. Further, the very identical epitopes were removed to improve the performance. The tool has accuracy approximately from 54–86% using various features such as dipeptide composition, binary profile, amino acid pair profile [45].

2.10 DiscoTope

The DiscoTope server uses three-dimensional protein structures to anticipate discontinuous B-cell epitopes. Surface accessibility (measured in terms of contact counts) and a unique epitope propensity amino acid score are used in the method. The final scores are computed by adding the propensity scores of nearby residues and the contact numbers [46].

DiscoTope detects 15.5 percent of residues in discontinuous epitopes with a 95 percent specificity. The predictions can guide experimental epitope mapping in both rational vaccine design and the development of diagnostic tools, potentially leading to more efficient epitope identification [47].

2.11 ElliPro

ElliPro predicts linear and discontinuous antibody epitopes based on the 3D structure of a protein antigen. ElliPro accepts protein structures in PDB format as input. If the input is a protein sequence, please go to methods for modeling and docking of antibody and protein 3D structures for more information on these methods. Thornton's method is implemented as a web platform that allows the prediction and visualization of antibody epitopes in a protein sequence or structure using a residue clustering algorithm, the MODELER program, and the Jmol viewer. ElliPro is based on the geometrical features of protein structure and requires no training. It could be used to predict many forms of protein–protein interactions.

When compared to DiscoTope, which is based on training datasets, ElliPro uses epitope features like residue solvent accessibility, amino acid propensities, inter-molecular contacts and spatial distribution of epitopes, thus improving the prediction ability [48].

2.12 EpiPred

EpiPred is a program that predicts structural epitopes unique to a given antibody. Epitope predictions from EpiPred can be utilized to increase antibody–antigen docking performance. The approach can be utilized using an antibody homology model as input.

Patches on the antigen structure are prioritized based on their likelihood of being the epitope. The program rescores the global docking findings of two rigid-body docking algorithms: ZDOCK and ClusPro, using epitope predictions.

Other approaches, such as DiscoTope or PEPITO annotate broad immunogenic/epitope like regions on the antigen without requiring any antibody information on input unlike epiPred [49].

The use of *in silico* models can significantly reduce the time and effort required to carry out an epitope discovery experiment. These methods include semi-automatic approaches for epitope discovery and incorporate high-throughput experimental tests for detecting MHC–peptide binding affinities. The *in silico* methods save a significant amount of resources in terms of both peptide binding classification accuracy and detecting immunogenic peptides. A competent immunologists' interpretation is required to appropriately confirm the outcome of such a prediction system [50].

3. Future scope

The discovery of vaccines is one of the most important aspects of world public health. Traditional vaccine design procedures have significant shortcomings, but the use of computational tools will overcome these limits. Because immunoinformatics approaches are more useful, modern technologies such as reverse vaccinology, epitope prediction, and structural vaccinology, as well as rational approaches, are in higher demand to produce new vaccine candidates.

The chapter describes the various bioinformatic tools that are available for determining immunogenic characteristics, locating T- and B-cell epitopes, and *in silico* technologies that are utilized in vaccine development.

4. Conclusion

The application of bioinformatic techniques has greatly accelerated the discovery of new medicinal targets in the post-genomic age. The availability of pathogenic microbe genome sequences has resulted in an increase in the discovery of genes and proteins that could be used to develop drugs or vaccines. Bioinformatic methods have been critical in the analysis of genome and protein sequences to uncover immunogenic proteins among organism's repertoires. Immunogenicity prediction methods are automated, and the entire proteome may be evaluated to identify top candidates with immunity-inducing features. Not only immunogenic proteins are been identified, but individual epitope mapping has also been completed. Methods for locating T- and B-cell epitopes are now available, which could lead to the development of epitope-based vaccinations.

Author details

Shilpa Shiragannavar* and Shivakumar Madagi
Department of Research and Studies in Bioinformatics and Biotechnology,
Karnataka State Akkamahadevi Women's University, Vijayapura, India

*Address all correspondence to: shilpa.shiragannavar@gmail.com

IntechOpen

References

[1] Naidoo N, Pawitan Y, Soong R, Cooper DN, Ku CS. Human genetics and genomics a decade after the release of the draft sequence of the human genome. Human Genomics. 2011;**5**(6):1-46

[2] Oli AN, Obialor WO, Ifeanyichukwu MO, Odimegwu DC, Okoyeh JN, Emechebe GO, et al. Immunoinformatics and vaccine development: an overview. Immuno-Targets and Therapy. 2020;**9**:13

[3] Gibas C, Jambeck P, Fenton J. Developing Bioinformatics Computer Skills. Sebastopol, USA: O'Reilly Media, Inc.; 2001. ISBN: 9781565926646

[4] Singh H. Bioinformatics: Benefits to Mankind. International Journal of PharmTech Research. 2016;**9**(4):242-248

[5] Yang Z, Bogdan P, Nazarian S. An in silico deep learning approach to multi-epitope vaccine design: a SARS-CoV-2 case study. Scientific Reports. 2021;**11**(1):1-21

[6] Chukwudozie OS, Gray CM, Fagbayi TA, Chukwuanukwu RC, Oyebanji VO, Bankole TT, et al. Immuno-informatics design of a multimeric epitope peptide based vaccine targeting SARS-CoV-2 spike glycoprotein. PLoS One. 2021;**16**(3):e0248061

[7] Yang Z, Bogdan P, Nazarian S. An in silico deep learning approach to multi-epitope vaccine design: a SARS-CoV-2 case study. Scientific Reports. 2021;**11**(1):1-21

[8] Doytchinova IA, Flower DR. VaxiJen: a server for prediction of protective antigens, tumour antigens and subunit vaccines. BMC Bioinformatics. 2007;**8**(1):1-7

[9] Doytchinova IA, Flower DR. Bioinformatic approach for identifying parasite and fungal candidate subunit vaccines. Open Vaccine Journal. 2008;**1**(1):4

[10] Dalsass M, Brozzi A, Medini D, Rappuoli R. Comparison of open-source reverse vaccinology programs for bacterial vaccine antigen discovery. Frontiers in Immunology. 2019;**10**:113

[11] Magnan CN, Zeller M, Kayala MA, Vigil A, Randall A, Felgner PL, et al. High-throughput prediction of protein antigenicity using protein microarray data. Bioinformatics. 2010;**26**(23): 2936-2943

[12] Rasheed MA, Raza S, Zohaib A, Riaz MI, Amin A, Awais M, et al. Immunoinformatics based prediction of recombinant multi-epitope vaccine for the control and prevention of SARS-CoV-2. Alexandria Engineering Journal. 2021;**60**(3):3087-3097

[13] Dimitrov I, Bangov I, Flower DR, Doytchinova I. AllerTOP v. 2—a server for in silico prediction of allergens. Journal of Molecular Modeling. 2014; **20**(6):1-6

[14] Shen H, Chou KC. Using optimized evidence-theoretic K-nearest neighbor classifier and pseudo-amino acid composition to predict membrane protein types. Biochemical and Bio-physical Research Communications. 2005;**334**(1):288-292

[15] Sanchez-Trincado JL, Gomez-Perosanz M, Reche PA. Fundamentals and methods for T-and B-cell epitope prediction. Journal of Immunology Research. 2017;**2017**: 2680160. DOI: 10.1155/2017/2680160. Epub 2017 Dec 28. PMID: 29445754; PMCID: PMC5763123.

[16] Adhikari A, Simha MV, Singh V, Jha RK, Upadhyay H. A Review on Immunosuppressive Drugs of Organ Transplantation. Dec 2019;**22**(14)

[17] Jespersen MC, Mahajan S, Peters B, Nielsen M, Marcatili P. Antibody specific B-cell epitope predictions: leveraging information from antibody-antigen protein complexes. Frontiers in Immunology. 2019;**10**:298

[18] Zobayer N, Hossain AA, Rahman MA. A combined view of B-cell epitope features in antigens. Bioinformation. 2019;**15**(7):530

[19] Delves PJ, Roitt IM. Encyclopedia of Immunology. The National Agricultural Library. 2nd ed. United States: Academic Press; 1998

[20] Alberts B, Johnson A, Lewis J, et al. Molecular Biology of the Cell. 4th edition. New York: Garland Science; 2002. T Cells and MHC Proteins. Available from: https://www.ncbi.nlm.nih.gov/books/NBK26926/

[21] Fleri W, Paul S, Dhanda SK, Mahajan S, Xu X, Peters B, et al. The immune epitope database and analysis resource in epitope discovery and synthetic vaccine design. Frontiers in Immunology. 2017;**8**:278

[22] Vita R, Mahajan S, Overton JA, Dhanda SK, Martini S, Cantrell JR, et al. The immune epitope database (IEDB): 2018 update. Nucleic Acids Research. 2019;**47**(D1):D339-D343

[23] Andreatta M, Nielsen M. Gapped sequence alignment using artificial neural networks: application to the MHC class I system. Bioinformatics. 2016;**32**(4):511-517

[24] Jurtz V, Paul S, Andreatta M, Marcatili P, Peters B, Nielsen M. NetMHCpan-4.0: improved peptide–MHC class I interaction predictions integrating eluted ligand and peptide binding affinity data. The Journal of Immunology. 2017;**199**(9):3360-3368

[25] Rammensee HG, Bachmann J, Emmerich NP, Bachor OA, Stevanović SS. SYFPEITHI: database for MHC ligands and peptide motifs. Immunogenetics. 1999;**50**(3):213-219

[26] Singh H, Raghava GP. ProPred1: prediction of promiscuous MHC Class-I binding sites. Bioinformatics. 2003; **19**(8):1009-1014

[27] Reche PA, Glutting JP, Reinherz EL. Prediction of MHC class I binding peptides using profile motifs. Human Immunology. 2002;**63**(9):701-709

[28] Doytchinova IA, Guan P, Flower DR. EpiJen: a server for multistep T cell epitope prediction. BMC Bioinformatics. 2006;7(1):1-1

[29] Guan P, Doytchinova IA, Zygouri C, Flower DR. MHCPred: a server for quantitative prediction of peptide–MHC binding. Nucleic Acids Research. 2003;**31**(13):3621-3624

[30] Zhang GL, Lin HH, Keskin DB, Reinherz EL, Brusic V. Dana-Farber repository for machine learning in immunology. Journal of Immunological Methods. 2011;**374**(1-2):18-25

[31] Jensen KK, Andreatta M, Marcatili P, Buus S, Greenbaum JA, Yan Z, et al. Improved methods for predicting peptide binding affinity to MHC class II molecules. Immunology. 2018;**154**(3):394-406

[32] Lata S, Bhasin M, Raghava GP. Application of machine learning techniques in predicting MHC binders. In: Immunoinformatics. Humana Press; 2007. pp. 201-215

[33] Terry FE, Moise L, Martin RF, et al. Time for T? Immunoinformatics addresses vaccine design for neglected tropical and emerging infectious diseases. Expert Review of Vaccines. 2015;**14**(1):21-35. DOI: 10.1586/14760584.2015.955478

[34] Moise L, McMurry JA, Buus S, Frey S, Martin WD, De Groot AS. In silico-accelerated identification of

conserved and immunogenic variola/
vaccinia T-cell epitopes. Vaccine.
2009;**27**(46):6471-6479. DOI: 10.1016/j.
vaccine.2009.06.018

[35] Pisitkun T, Hoffert JD, Saeed F,
Knepper MA. NHLBI-AbDesigner: an
online tool for design of peptide-directed
antibodies. The American Journal of
Physiology. 2012;**302**(1):C154-C164.
DOI: 10.1152/ajpcell.00325.2011

[36] Vivona S, Filippo B, Francesco F.
NERVE: new Enhanced reverse
vaccinology environment. BMC
Biotechnology. 2006;**6**:35. DOI:
10.1186/1472-6750-6-35

[37] Moise L, Gutierrez A, Kibria F, et al.
iVAX: an integrated toolkit for the
selection and optimization of antigens
and the design of epitope-driven
vaccines. Human Vaccines &
Immunotherapeutics. 2015;**11**(9):2312-
2321. DOI: 10.1080/21645515.
2015.1061159

[38] De Groot AS, Bosma A, Chinai N,
et al. From genome to vaccine: in silico
predictions, ex vivo verification.
Vaccine. 2015;**19**(31):4385-4395. DOI:
10.1016/S0264-410X(01)00145-1

[39] Soria-Guerra RE, Nieto-Gomez R,
Govea-Alonso DO, Rosales-Mendoza S.
An overview of bioinformatics tools for
epitope prediction: implications on
vaccine development. Journal of
Biomedical Informatics. 2015;**53**:405-
414. DOI: 10.1016/j.jbi.2014.11.003

[40] Bui HH, Sidney J, Dinh K,
Southwood S, Newman MJ, Sette A.
Predicting population coverage of T-cell
epitope-based diagnostics and vaccines.
BMC Bioinformatics. 2006;**17**:153

[41] EL-Manzalawy Y, Dobbs D,
Honavar V. Predicting linear B-cell
epitopes using string kernels. Journal of
Molecular Recognition: An Interdisciplinary Journal. 2008;**21**(4):243-
255. DOI: 10.1002/jmr.893. PMID:
18496882; PMCID: PMC2683948.

[42] Liu T, Shi K, Li W. Deep learning
methods improve linear B-cell epitope
prediction. BioData Mining. 2020;**13**(1):
1-3

[43] Saha S, Raghava GPS. Prediction of
Continuous B-cell Epitopes in an Antigen
Using Recurrent Neural Network.
Proteins. 2006;**65**(1):40, 16894596-48

[44] Jespersen MC, Peters B, Nielsen M,
Marcatili P. BepiPred-2.0: improving
sequence-based B-cell epitope prediction
using conformational epitopes. Nucleic
Acids Research. 2017;**45**(W1):W24-W29

[45] Singh H, Ansari HR, Raghava GP.
Improved method for linear B-cell epitope
prediction using antigen's primary
sequence. PLoS One. 2013;**8**(5):e62216

[46] Kringelum JV, Lundegaard C,
Lund O, Nielsen M. Reliable B cell
epitope predictions: impacts of method
development and improved
benchmarking. PLoS Computational
Biology. 2012;**8**(12):e1002829

[47] Haste Andersen P, Nielsen M,
Lund OL. Prediction of residues in
discontinuous B-cell epitopes using
protein 3D structures. Protein Science.
2006;**15**(11):2558-2567

[48] Ponomarenko J, Bui HH, Li W,
Fusseder N, Bourne PE, Sette A, et al.
ElliPro: a new structure-based tool for
the prediction of antibody epitopes.
BMC Bioinformatics. 2008;**9**(1):1-8

[49] Krawczyk K, Liu X, Baker T, Shi J,
Deane CM. Improving B-cell epitope
prediction and its application to global
antibody-antigen docking.
Bioinformatics. 2014;**30**(16):2288-2294

[50] Lundegaard C, Lund O, Buus S,
Nielsen M. Major histocompatibility
complex class I binding predictions as a
tool in epitope discovery. Immunology.
2010;**130**(3):309-318

Chapter 3

Perspective Chapter: Next-Generation Vaccines Based on Self-Amplifying RNA

Fatemeh Nafian, Simin Nafian, Ghazal Soleymani,
Zahra Pourmanouchehri, Mahnaz Kiyanjam,
Sharareh Berenji Jalaei, Hanie Jeyroudi
and Sayed Mohammad Mohammdi

Abstract

Recently, nucleic acid-based RNA and DNA vaccines have represented a better solution to avoid infectious diseases than "traditional" live and non-live vaccines. Synthetic RNA and DNA molecules allow scalable, rapid, and cell-free production of vaccines in response to an emerging disease such as the current COVID-19 pandemic. The development process begins with laboratory transcription of sequences encoding antigens, which are then formulated for delivery. The various potent of RNA over live and inactivated viruses are proven by advances in delivery approaches. These vaccines contain no infectious elements nor the risk of stable integration with the host cell genome compared to conventional vaccines. Conventional mRNA-based vaccines transfer genes of interest (GOI) of attenuated mRNA viruses to individual host cells. Synthetic mRNA in liposomes forms a modern, refined sample, resulting in a safer version of live attenuated RNA viruses. Self-amplifying RNA (saRNA) is a replicating version of mRNA-based vaccines that encode both (GOI) and viral replication machinery. saRNA is required at lower doses than conventional mRNA, which may improve immunization. Here we provide an overview of current mRNA vaccine approaches, summarize highlight challenges and recent successes, and offer perspectives on the future of mRNA vaccines.

Keywords: vaccine (s), self-amplifying RNA (saRNA), in vitro transcription (IVT), design of experiments, nucleic acid, messenger RNA (mRNA), innate immune stimulation

1. Introduction

Vaccines have been the most successful biomedical invention to prevent the morbidity and mortality caused by infectious diseases [1]. A vaccine stimulates the immune system to produce antibodies against target antigens, preventing infection, reducing disease severity, and decreasing the rate of hospitalization. Early vaccines were based on live, non-live (inactivated), or attenuated replicating strains of the relevant pathogenic organism from those that had only segments of a pathogen or killed whole organisms. In the second half of the last century, the development of the industrial production of a new series of vaccines was known as the advancing years of vaccinology.

IntechOpen

It started by generating vaccines against Rubella, Mumps, and Measles in the 1960s, which was extended with the production of the Chickenpox vaccine, and deactivating Japanese encephalitis. The induction of a protective immune response could be a target for new advanced vaccines. Cultivation techniques under dominated conditions have been used to process mass vaccine production. In the 1980s, "conjugate vaccines" were used to stimulate immune responses against capsular polysaccharides and proteins of pathogens. Polysaccharide antigens can also cause protective immune responses and are the basis of vaccines that have been evolved to prevent several bacterial infections, such as pneumonia and meningitis caused by Streptococcus pneumonia, since the late 1980s. Conjugate vaccines were presented either in the form of whole or inactivated pathogens or as structural parts. To gain complete prevent, the vaccine must contain antigens that are either led by the pathogen or produced synthetically to represent the segment of the pathogen. The main component of most vaccines is one or more protein antigens that breed immune responses that create protection. Therefore, "recombinant vaccines" were advanced using genetic engineering to produce multivalent vaccines and balance the efficiency of the immune response and the safety of antigens for immunogenicity.

2. Nucleic acid-based vaccines

In 1990, vaccine development came into its golden age with the introduction of nucleic acid-based vaccines, including viral vectors, plasmid DNA (pDNA), and

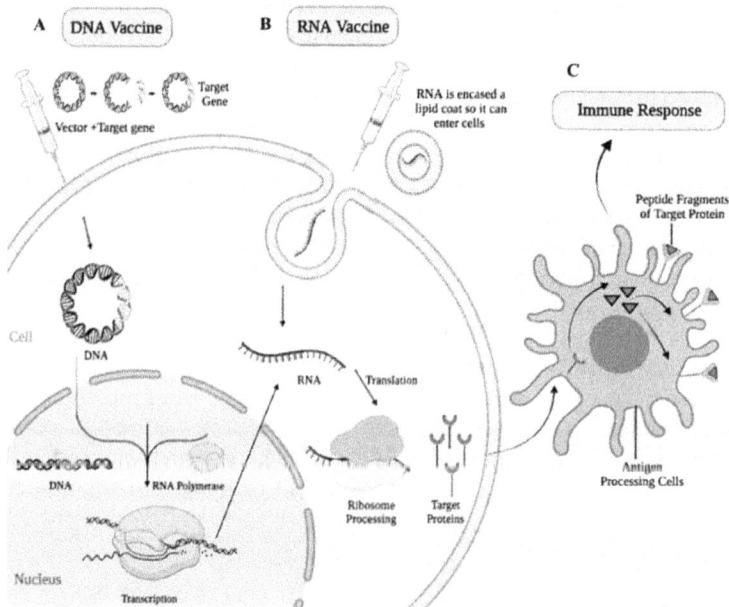

Figure 1.
Molecular basis for the immunostimulatory activity of next-generation nucleic acid vaccines. A) In the DNA vaccine, the target antigen is inserted into a vector. After DNA vaccine injection, the inserted antigen must cross the cell and nuclear membranes to use cellular enzymes which allows the antigen transcripts into mRNAs in the nuclear and then translates into immunogen proteins in the cytoplasm. B) In RNA vaccines the RNA encoding the immunogen protein is formulated into nanoparticles to be delivered into the cell, then its endocytosis and mRNA are releases into the cytoplasm. By entering mRNA into the cytoplasm, the RNA translates to immunogenic proteins. C) the produced protein will be presented upon the surface of the cell to trigger an immune response through antigen-presenting cells (APCs), such as macrophages and dendritic cells. APCs process proteins, break them into peptides and present them in conjunction with MHC molecules on the cell surface where they may interact with appropriate T cell receptors. This figure was created using BioRender (http://www.biorender.com).

mRNA as safer alternatives. They consist of one or more genes of interest (GOI) encoding practically active antigens from an addressed pathogen. After intramuscular injection of RNA or DNA vectors, they are up-taken by immune effector cells and host-arbitrated expressed, resulting in induction of both cellular and humoral immunity (**Figure 1**) [2]. It has been demonstrated that they can induce broadly protective immune responses with a safe approach against infectious and non-infectious diseases. However, nucleic acid vaccines are basically therapeutic agents for cancer. The main challenge is developing ways to prevent or treat infectious diseases such as COVID-19 and human immunodeficiency virus (HIV). With the advent of nucleic acid vaccines, the time and cost of vaccine design and production have been considerably cut, since once the platform has been established for GOI synthesis and insertion into an appropriate expression vector [3]. Accordingly, it might be useful for the development of vaccines against emerging pandemic infectious diseases [4]. Upon vaccination, they mimic a viral infection to express antigens in situ and lower doses are required to stimulate both humoral and cellular responses [5]. RNA vaccines have some advantages over DNA vaccines, such as being able to enter non-dividing cells to the cytosolic expression of proteins. In contrast, using DNA shows the perceived risk of integration into the host genome, since it uptakes into cells and enters into the nucleus due to the breakdown of the nuclear membrane during cell division. However, there are difficulties in producing the quantities of mRNA required to be produced in vivo [4]. A "vaccine on-demand" approach can provide a rapid research and development process, large-scale production, and distribution for nucleic acid based-vaccines [6].

3. RNA vaccines

Over the past years, mRNA has provided a promising technology in the field of vaccine development with several beneficial features overkilled, live attenuated viruses, and subunit as well as DNA vaccines [7]. First, mRNA is a safe platform with no potential risk of infection or genomic integration. mRNA half-life can be regulated in vivo using various modifications and delivery methods [8]. The mRNA immunogenicity can be down-modulated to further increase the safety profile [9]. Second, various modifications can increase efficacy, stability, and expression levels of mRNA [10]. In vivo delivery can be well-organized by formulating mRNA into carrier molecules, allowing rapid uptake and expression in the cytoplasm [11]. Third, mRNA vaccines can be manufactured rapidly and inexpensively on a large scale with a high yield of IVT reactions.

The mRNA vaccine falls into two basic types: conventional non-replicating mRNA and self-amplifying RNA (saRNA). Both approaches show essential elements of a eukaryotic mRNA: a cap structure [m7Gp3N], a 5′ UTR, an open reading frame (ORF) encoding the gene of interest (GOI), a 3′ UTR, and a tail of 40–120 adenosine residues [poly (A) tail] (**Figure 2**) [12]. Self-amplifying mRNA vaccines are derived from the engineered RNA genomes of plus-strand RNA viruses such as alphaviruses or flaviviruses, and picornaviruses [13, 14]. Therefore, it encodes not only the antigen of interest flanked by 5′ and 3′ UTRs but also contains an amplicon required for intracellular RNA amplification enhancing antigen expression levels [13, 15, 16].

Both groups can be produced in vitro transcription of mRNA (IVT mRNA), as a cell-free system, using an enzymatic transcription reaction from a linearized pDNA template [17]. The RNA manufacturing begins with the construction of the pDNA molecule that is used as a template for an IVT mRNA using a promoter with a high binding affinity for a DNA-dependent RNA polymerase, and a restriction site for

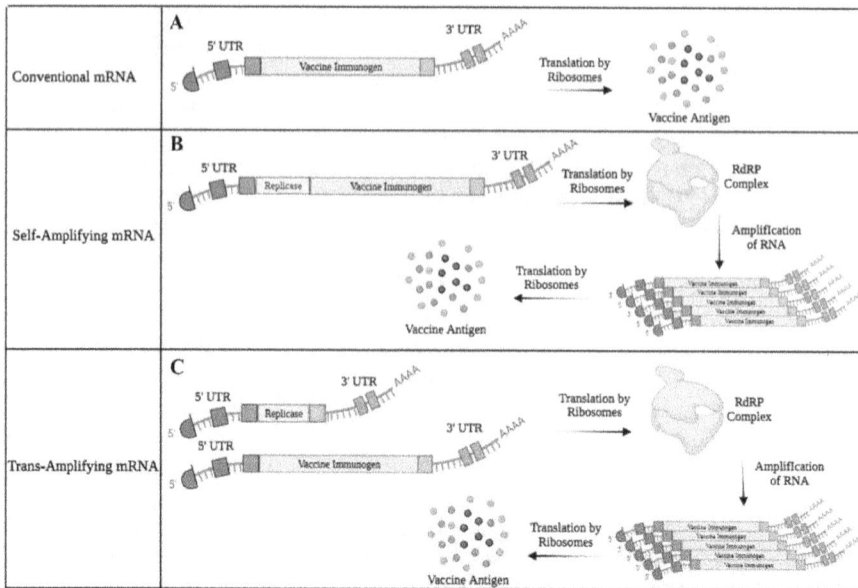

Figure 2.
Schematic of different parts of conventional mRNA, self-amplifying mRNA, and trans-amplifying mRNA vaccines. A) In conventional mRNA, the essential elements are presented, included: a cap structure, a 5′ UTR, an ORF encoding immunogenic protein, a 3′ UTR, and a 3′poly (A) tail, and also the mRNA translates by ribosomes. B) self-amplifying mRNA contains a cap structure, a 5′ UTR, an ORF encoding immunogenic protein, a 3′ UTR, and a 3′poly (A) tail and extra region called replicase. Replicase is a replicating polyprotein complex which this constructs from capping enzyme, helicase, poly (A) polymerase, and an RNA-dependent RNA-polymerase (RdRP) which causes high expression of downstream antigens and increasing immunogenicity. First, the mRNA translates by ribosomes to assemble the RdRp complex to amplify the mRNA then in the next step these amplified mRNAs translate to the vaccine target protein. C) In trans-amplifying mRNA, the replicase and ORF encoding immunogenic protein is separated into two mRNA, but co-delivered to the target cells, which both have a cap structure, a 5′ UTR, a 3′ UTR, and a 3′poly (A) tail. First of the replicase translates to the RdRp complex then utilizes it to amplify. At the end mRNAs, containing ORF encoding immunogenic protein will be translated by ribosomes. This figure was created using BioRender (http://www.biorender.com).

insertion of the specific sequence encoding the target antigen. The ORF must be inserted without affecting the overall physicochemical characteristics of the mRNA molecule. IVT mRNA occurs in three steps of transcription; initiation, elongation, and termination with the help of a linearized plasmid DNA that has a promoter, NTPs, RNA polymerase, and mg2+. The enzyme elongates the RNA transcript until it runs off the end of the template. Then, pDNA is degraded by incubation with DNase, and a cap or a synthetic cap analog is enzymatically added to the 5′ end of the mRNA [18–20]. A co-transcriptional capping strategy was also developed to add a natural 5′ cap structure to a specific start sequence during IVT [21]. This strategy results in innate immune activation when the IVT mRNA is prepared [22, 23]. The presence of a 5′cap structure protects mRNA from intracellular nuclease digestion and is also crucial for efficient translation in vivo [24, 25]. In the end, the mRNA is purified to remove reaction agents, including residual pDNA, enzymes, truncated or double-stranded transcripts [26, 27]. There are several factors that must be addressed before using mRNA, including the ability to express sufficient levels of antigen, immunogenicity, stability, and toxicity of formulated RNA, as well as the possibility of having negative consequences on unexpected and undesired tissues [28]. The establishment of valuable guidance for rapid, simple, and inexpensive mass production of mRNA is a critical requirement for the future implementation of mRNA vaccines [29].

Once the purified mRNA enters the cytosol, the cellular translation and post-translation machinery produce a properly folded, fully functional protein. IVT mRNA is finally degraded by normal physiological processes, thus reducing the risk of metabolite toxicity. In some cases, it has validated the immunogenicity of various mRNA vaccine platforms in recent years [30–33]. Engineering of the RNA sequence has increased translation and prolonged antigen expression in vivo. The first successful vaccination by the IVT mRNA was reported around three decades ago when an intramuscular injection in mice resulted in the local production of an encoded reporter protein and generalization of immune responses against the antigen [34]. However, the early results led to no substantial investment in developing mRNA vaccines, due to concerns about mRNA instability, inefficient in vivo delivery, and high innate immunogenicity. Instead, the field pursued DNA-based and protein-based therapeutic approaches [35, 36].

3.1 Conventional mRNA vaccines

A conventional mRNA vaccine only encodes the sequence of the specific antigen flanked by transcription regulatory regions. The major advantages of the conventional mRNA vaccine are the simplicity and small size of the RNA molecule. However, the stability and efficiency of conventional mRNA in vivo is limited and needs to be optimized in RNA structural elements and formulation approach [37]. The cap or its analogs, UTRs, and the poly (A) tail are crucial elements for stability, accessibility, and interaction with the translation machinery of the mRNA vaccine [38–41]. Codon usage is required to enhance protein expression from DNA, RNA, and viral vector vaccines [42, 43]. The nucleoside base of mRNA can be chemically modified coupled with chromatographic purification to remove dsRNA contaminants, which are improper immune-stimulatory [44, 45]. Although activation of innate immunity is required for vaccination, its excessive activation interferes with antigen production and adaptive immunity [46, 47]. When modified mRNA is highly purified, the highest levels of protein expression and immunogenicity are observed [48].

3.2 Self-amplifying RNA (saRNA) vaccines

Self-amplifying RNA (also called replicon RNA) is one of the most immune-responsive types of mRNA since it activates several Toll-like receptors (TLRs) to generate very strong immune responses. saRNA not only has the basic parts of eukaryotic mRNA (a cap, an ORF, a sub-genomic promoter, a poly (A) tail, 3' and 5' UTR flanks) but also encodes a replicating polyprotein complex including an RNA-dependent RNA-Polymerase, capping, helicase, and poly (A) polymerase (**Figure 2**). The replicative features of positive-stranded RNA viruses are mimicked to highly express the antigen and increase immunogenicity. The gene of interest is placed downstream of the replicon construct, which is under the control of the promoter. Entering saRNA into the cell cytoplasm immediately couples with the translation of a replicase complex that recognizes a subgenomic promoter and amplifies a smaller mRNA (subgenomic RNA). The most predominant antigens in saRNA are viral glycoproteins, although this has recently been expanded to include the proteins of bacterial infections, parasites, and cancer. A more novel saRNA encodes monoclonal antibodies for passive vaccination. If necessary to encode multiple antigens, it can be advantageous to use separate saRNA constructs since the pDNA construct does have size limitations [49]. saRNA vaccines against bacterial antigens are limited to protein targets, as opposed to polysaccharides and non-protein surface markers.

The innate immune system has advanced to recognize pathogen-associated molecular patterns (PAMPs) via binding pattern recognition receptors (PRRs) with them. PRRs can be represented as cytosolic receptors including nucleotide-like receptors (NLRs) and RIG-I like receptors (RLRs) or as endosomal toll-like receptors (TLRs) [50, 51]. Innate immune activation can recognize both conventional mRNA and saRNA and arrange signaling pathways to strongly battle the pathogen [52]. saRNA is more immunogenic than conventional mRNA. The optimization of both translation and purity of mRNA vaccines (conventional mRNA or saRNA) can be used to overcome the problems of immune-stimulating activities. Some discoveries emphasize the capacity to synthesize high quality and quantity of mRNA through IVT [53]. The full-length mRNA molecule can be produced by IVT from a pDNA template and delivered as either synthetically unformulated RNA, or as formulated into nanoparticles if structural genes are provided in trans. Alternatively, saRNA can also be produced directly in vivo, by delivering a pDNA containing the replicon complex and GOI into the target cells [54]. The efficacy of nanoparticles as a vaccine carrier must be engineered to correctly condense processed heterologous proteins and deliver antigen to specific cell types, such as APCs [54, 55]. Activation of dendritic cells by mRNA nanoparticles results in a wave of cytokine cascade and subsequently a vaccine-elicited adaptive immune response [56]. One of the challenges with mRNA vaccines is the determination of sufficient quantities of RNA sequence, integrity, and purity for use in the target population (**Figure 3**) [57–59].

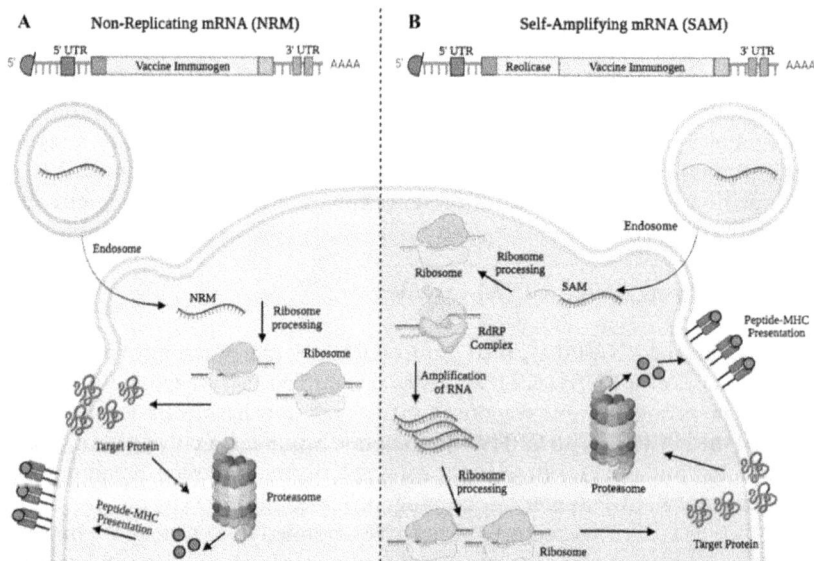

Figure 3.
Schematic of different protein production of conventional mRNA and self-amplifying mRNA in APCs. Both mRNAs (A and B) can be encapsulated in nanoparticles (NPs) to preserve them from degradation and facilitate cellular uptake. Membrane-derived endocytic pathways are commonly used for cellular uptake of mRNA with its delivery system. A) after entry of SAM into cytoplasm the translation begins to build up the RdRp complex and in the next step the RdRp complex amplifies the mRNAs. mRNAs translate by ribosomes and then some of them will be delivered directly on the surface and others will be degraded by proteasome which leads to MHC presentation. B) In NRM, the mRNA is just translated and does not have replication. Like to SAM the numbers of the target proteins are expressed on the surface directly on the rest of them through MHC presentation. Once the formulated mRNA enters the cytosol, it directly translates and post-translates into a fully functional antigen to be ready for MHC presentation on the cell surface. The presented protein can induce both innate and adaptive immune responses. Furthermore, self-amplifying mRNA encodes the replicating complex that is required for intracellular RNA amplification. This figure was created using BioRender. (http://www.biorender.com).

4. Delivery systems for mRNA vaccines

Naked mRNA is quickly degraded by extracellular RNases and is not internalized efficiently. Anionic saRNA constructs are relatively large (9000 to 15,000 n) and need to be condensed by a cationic carrier into a nanoparticle of ~100 nm in size that encourages uptake into target cells and protects the saRNA from degradation [60]. A great variety of in vitro and in vivo transfection reagents can facilitate cellular uptake of mRNA and protect it from degradation. Unmodified RNA has limited stability in the bloodstream and passes through the cell membrane and immunogenicity. RNA is a single hydrophilic molecule and is likely to form a secondary or territory structure, leading to an unfavorable problem with delivery approaches. The vectors used for effective cell delivery of mRNA vaccines include viral vectors such as adenoviral, adeno-associated viral, retroviral, and lentiviral vectors, as well as nanoparticles made of lipids, polymers, and inorganic compounds. Viral vectors present high-efficiency transfection by design through infection gene deletion, viral replication, and assembly. On the other hand, nanoparticle (NPs) vectors offer advantages such as greater safety, flexible administration, wider adoption, and unlimited transgene size and may represent the future of next-generation vaccines.

The early delivery system was a combination of minor arginine-rich cationic proteins, protamine, and mRNA. Protamine-complexed mRNA reduced protein expression. That's why a mixture of free and protamine-complexed mRNA is used [61]. NPs offer a variety of biomaterial alternatives to formulate mRNA, facilitate cell internalization, increase specific immune cell targeting through surface modifications, and boost endosomal escape using pH-sensitive materials [62]. NPs, as a strong adjuvant, enhance protection through synergistic effects since they allow for the generation of cocktail vaccines in a single particle and for the delivery of numerous nucleic acids to the same target cell [62]. There has been new research into liposomal, polymeric, inorganic, lipidoid, and peptide-based nanoparticles, with a wide range of different 3D structures, sizes, and modifications, all aimed at increasing the effectiveness of nucleic acid delivery. The recent systems for mRNA vaccine delivery are liposome nanoparticles (LNP), which are lipid bilayer-coated artificial vesicles. Cationic lipids bind to negatively charged nucleic acids through their positive-charged hydrophilic heads, and the hydrophobic lipid tails encapsulate them. On the other hand, neutral lipids can be used to improve transfection efficiency and stability. The quantity of charged groups per molecule, geometric shape, and type of the lipid anchor are all important factors in transfection efficiency [62]. The LNP-encapsulated saRNAs induce higher antibody titers, pathogen neutralization (IC50), and antigen-specific CD4+ and CD8+ T cell responses than electroporated pDNA [62]. Recently, the combination of LNPs with dendritic cell (DC) targeting, UTR optimization, lipopolymer design, and ionizable lipids has advanced the field of saRNA vaccination. However, there are concerns about the toxicity of cationic LNPs, caused by membrane disruption, as well as endosomal escapes that interact with LNPs.

Polymeric nanoparticles (PNPs) that are also used for mRNA vaccine delivery are typically made from biocompatible and biodegradable polymers. They have a wide range of physicochemical properties, which may be structurally modified for regulated release of the gene contents [62]. Cationic polymers encapsulate mRNA via electrostatic interactions to generate polymer-mRNA polyplexes. Polyamidoamine dendrimers (PAMAM) and polyethylenimine (PEI) are two commonly used polymers. PNPs preserve mRNAs and also facilitate their entrance into cells. PEI offers additional benefits, including a higher protonation ratio of amine groups and a high buffer capacity over a wide pH range [62]. PAMAM is a biocompatible and highly branched cationic polymer that enables functionalization to deliver multiple antigen-expressing replicons at once. The low transfection effectiveness and cytotoxicity of PNPs-mRNA delivery are remaining challenges.

For nucleic acid delivery, inorganic NPs have been extensively studied. Inorganic NPs have a lower size than polymeric/liposomal NPs, a limited size distribution, and ligand conjugation-friendly surface chemistry. Gold nanoparticles (AuNPs) have a wide range of electromagnetic properties required for mRNA delivery and their surfaces are easily modified with a variety of ligands. Mesoporous silica nanoparticles (MSNs) are another form of biodegradable inorganic NPs that have very porous nanostructures. The porosity can provide a large surface area for chemical modification and mRNA encapsulation in multiple targeting carriers. Positively charged peptides can form natural NPs for mRNA vaccine delivery due to the presence of lysine and arginine residues. Cell-penetrating peptides can enhance delivery efficacy by creating complexes with nucleic acids. Furthermore, virus-like particles (VLPs) can be used to deliver nucleic acid vaccines, although they may be cleared by phagocytes. Nowadays, next-generation nucleic acid vaccines have been focused on developing transfection effectiveness, optimizing the safety profiles of NP formulations, inducing protective immune responses, and using mixtures of nucleic acid vaccines to target the same immune cell of interest in vivo.

5. Conclusions

Vaccinology is moving toward artificial polymer platforms that allow for fast, scalable, and non-cellular mass production of vaccines. Using mRNA vaccines, there is no chance for undesirable mutations, including insertion, breakage, frameshift, or rearrangements, caused by genome integration [63, 64]. Both conventional and self-amplifying RNA vaccines can be simply designed without restrictions on the size and sequence of antigens. Additionally, multiple antigens can be applied downstream of a robust promoter to have powerful transcription and translation, thus lower doses will be used. However, this chapter demonstrated how self-amplifying RNAs are more potent than conventional types due to productive amplification of mRNA directly within the cytoplasm, the adaptability of applying delivery vectors, and induction of both humoral and cellular immunity for powerful and long-lasting prevention against chronic infectious diseases. Nucleic acid vaccines are safer than infective agents and certain preclinical safety studies may not be necessary, which would further shorten development time and cost. The mRNA vaccines do not require preservation in a cold chain because they are not a live infective vector. All that is needed for mRNA vaccines is to obtain gene sequence information to manufacture an optimized pDNA template using completely synthetic processes or clone it into appropriate expression vectors. They can be delivered for antigen expression in situ without the need to cross the nuclear membrane barrier for protein expression and can express complex antigens without packaging constraints. Finally, it is important to note that both types of nucleic acid-based vaccines have significant advantages over conventional vaccines and thus could be ideal for rapid responses to newly emerging pathogens.

Acknowledgements

We are thankful to Dr. Kamali Doust Azad and Dr. Bagherian for technical supports.

Conflict of interest

The authors declare no conflict of interest.

Notes

All authors whose names appear in the submission made substantial contributions to the conception, design, and acquisition of data.

Abbreviations

saRNA	self-amplifying RNA
taRNA	trans-amplifying RNA
IVT	in vitro transcription
GOI	genes of interest
pDNA	plasmid DNA
APC	antigen-presenting cell
DC	dendritic cell
ORF	open reading frame
TLR	Toll-like receptor
NLR	nucleotide like receptor
RLR	RIG-I like receptor
PAMP	pathogen-associated molecular pattern
PRR	pattern recognition receptor
RdRp	RNA-dependent RNA polymerase
NPs	nanoparticles
PNP	polymeric nanoparticles
AuNPs	gold nanoparticles
MSNs	mesoporous silica nanoparticles
PAMAM	polyamidoamine dendrimers
PEI	polyethylenimine

Author details

Fatemeh Nafian[1*], Simin Nafian[2], Ghazal Soleymani[3], Zahra Pourmanouchehri[4], Mahnaz Kiyanjam[5], Sharareh Berenji Jalaei[6], Hanie Jeyroudi[4] and Sayed Mohammad Mohammdi[4]

1 Department of Medical Laboratory Science, Tehran Medical Sciences, Islamic Azad University, Tehran, Iran

2 Department of Stem Cell and Regenerative Medicine, National Institute of Genetic Engineering and Biotechnology (NIGEB), Tehran, Iran

3 Department of Genetics, Islamic Azad University Tehran Medical Sciences, Tehran, Iran

4 Department of Cellular and Molecular Biology, Tehran Medical Sciences, Islamic Azad University, Tehran, Iran

5 Department of Biochemistry, Tehran Medical Sciences, Islamic Azad University, Tehran, Iran

6 Department of Microbiology, Tehran Medical Sciences, Islamic Azad University, Tehran, Iran

*Address all correspondence to: f.nafian@iautmu.ac.ir

IntechOpen

References

[1] van de Berg D, Kis Z, Behmer CF, Samnuan K, Blakney AK, Kontoravdi C, et al. Quality by design modelling to support rapid RNA vaccine production against emerging infectious diseases. npj Vaccines. 2021;**61**:1-10

[2] Tahamtan A, Charostad J, Hoseini Shokouh SJ, Barati M. An overview of history, evolution, and manufacturing of various generations of vaccines. Journal of Archives in Military Medicine. 2017;**53**:1-7

[3] Rauch S, Jasny E, Schmidt KE, Petsch B. New vaccine technologies to combat outbreak situations. Frontiers in Immunology. 1963;**2018**:9

[4] Naik R and Peden K. Regulatory Considerations on the Development of mRNA Vaccines. Berlin, Heidelberg: Springer; 2020

[5] Liu MA. Immunologic basis of vaccine vectors. Immunity. 2010;**334**:504-515

[6] Maruggi G, Zhang C, Li J, Ulmer JB, Yu D. mRNA as a transformative technology for vaccine development to control infectious diseases. Molecular Therapy. 2019;**274**:757-772

[7] Scorza FB, Pardi N. New kids on the block: RNA-based influenza virus vaccines. Vaccine. 2018;**62**:20

[8] Guan S, Rosenecker J. Nanotechnologies in delivery of mRNA therapeutics using nonviral vector-based delivery systems. Gene Therapy. 2017;**243**:133-143

[9] Weissman D. mRNA transcript therapy. Expert Review of Vaccines. 2015;**142**:265-281

[10] Thess A, Grund S, Mui BL, Hope MJ, Baumhof P, Fotin-Mleczek M, et al. Sequence-engineered mRNA without chemical nucleoside modifications enables an effective protein therapy in large animals. Molecular Therapy. 2015;**239**:1456-1464

[11] Kauffman KJ, Webber MJ, Anderson DG. Materials for non-viral intracellular delivery of messenger RNA therapeutics. Journal of Controlled Release. 2016;**240**:227-234

[12] Geall AJ, Mandl CW, Ulmer JB. RNA: The new revolution in nucleic acid vaccines. In: Seminars in Immunology. London, England: Elsevier Inc; 2013

[13] Tews BA, Meyers G. Self-replicating RNA. RNA Vaccines. 2017;**1499**:15-35

[14] Lundstrom K. Replicon RNA viral vectors as vaccines. Vaccine. 2016;**44**:39

[15] Brito LA, Kommareddy S, Maione D, Uematsu Y, Giovani C, Scorza FB, et al. Self-amplifying mRNA vaccines. Advances in Genetics. 2015;**89**:179-233

[16] Ulmer JB, Mason PW, Geall A, Mandl CW. RNA-based vaccines. Vaccine. 2012;**3030**:4414-4418

[17] Pardi N, Muramatsu H, Weissman D, Karikó K. In vitro transcription of long RNA containing modified nucleosides. In: Synthetic Messenger RNA and Cell Metabolism Modulation. Totowa, NJ: Humana Press; 2013. pp. 29-42

[18] Martin S, Moss B. Modification of RNA by mRNA guanylyltransferase and mRNA (guanine-7-) methyltransferase from vaccinia virions. Journal of Biological Chemistry. 1975;**25024**: 9330-9335

[19] Dwarki V, Malone RW, Verma IM. [43] Cationic liposome-mediated RNA

transfection. Methods in Enzymology. 1993;**217**:644-654

[20] Stepinski J, Waddell C, Stolarski R, Darzynkiewicz E, Rhoads RE. Synthesis and properties of mRNAs containing the novel "anti-reverse" cap analogs 7-methyl (3′-O-methyl) GpppG and 7-methyl (3′-deoxy) GpppG. RNA. 2001;**710**:1486-1495

[21] Vaidyanathan S, Azizian KT, Haque AA, Henderson JM, Hendel A, Shore S, et al. Uridine depletion and chemical modification increase Cas9 mRNA activity and reduce immunogenicity without HPLC purification. Molecular Therapy--Nucleic Acids. 2018;**12**:530-542

[22] Devarkar SC, Wang C, Miller MT, Ramanathan A, Jiang F, Khan AG, et al. Structural basis for m7G recognition and 2′-O-methyl discrimination in capped RNAs by the innate immune receptor RIG-I. Proceedings of the National Academy of Sciences. 2016;**1133**:596-601

[23] Schuberth-Wagner C, Ludwig J, Bruder AK, Herzner A-M, Zillinger T, Goldeck M, et al. A conserved histidine in the RNA sensor RIG-I controls immune tolerance to N1-2′ O-methylated self RNA. Immunity. 2015;**431**:41-51

[24] Li Y, Kiledjian M. Regulation of mRNA decapping. Wiley Interdisciplinary Reviews: RNA. 2010;**12**:253-265

[25] Marcotrigiano J, Gingras A-C, Sonenberg N, Burley SK. Cocrystal structure of the messenger RNA 5′ cap-binding protein (eIF4E) bound to 7-methyl-GDP. Cell. 1997;**896**:951-961

[26] Pardi N, Hogan MJ, Weissman D. Recent advances in mRNA vaccine technology. Current Opinion in Immunology. 2020;**65**:14-20

[27] Baiersdörfer M, Boros G, Muramatsu H, Mahiny A, Vlatkovic I, Sahin U, et al. A facile method for the removal of dsRNA contaminant from in vitro-transcribed mRNA. Molecular Therapy--Nucleic Acids. 2019;**15**:26-35

[28] Blakney A. The next generation of RNA vaccines: Self-amplifying RNA. The Biochemist. 2021;**43**:14-17

[29] Weissman D, Pardi N, Muramatsu H, Karikó K. HPLC purification of in vitro transcribed long RNA. In: Synthetic messenger RNA and cell metabolism modulation. Totowa, NJ: Humana Press, Springer; 2013. pp. 43-54

[30] Geall AJ, Verma A, Otten GR, Shaw CA, Hekele A, Banerjee K, et al. Nonviral delivery of self-amplifying RNA vaccines. Proceedings of the National Academy of Sciences. 2012;**10936**:14604-14609

[31] Pardi N, Tuyishime S, Muramatsu H, Kariko K, Mui BL, Tam YK, et al. Expression kinetics of nucleoside-modified mRNA delivered in lipid nanoparticles to mice by various routes. Journal of Controlled Release. 2015;**217**:345-351

[32] Bahl K, Senn JJ, Yuzhakov O, Bulychev A, Brito LA, Hassett KJ, et al. Preclinical and clinical demonstration of immunogenicity by mRNA vaccines against H10N8 and H7N9 influenza viruses. Molecular Therapy. 2017;**256**:1316-1327

[33] Pardi N, Hogan MJ, Pelc RS, Muramatsu H, Andersen H, DeMaso CR, et al. Zika virus protection by a single low-dose nucleoside-modified mRNA vaccination. Nature. 2017;**5437644**:248-251

[34] Wolff JA, Malone RW, Williams P, Chong W, Acsadi G, Jani A, et al. Direct gene transfer into mouse muscle in vivo. Science. 1990;**2474949**:1465-1468

[35] Suschak JJ, Williams JA, Schmaljohn CS. Advancements in DNA vaccine vectors, non-mechanical delivery methods, and molecular adjuvants to increase immunogenicity. Human Vaccines & Immuno therapeutics. 2017;**1312**:2837-2848

[36] Jones CH, Hakansson AP, Pfeifer BA. Biomaterials at the interface of nano-and micro-scale vector–cellular interactions in genetic vaccine design. Journal of Materials Chemistry B. 2014;**246**:8053-8068

[37] Ross J. mRNA stability in mammalian cells. Microbiological Reviews. 1995;**593**:423-450

[38] Andries O, Mc Cafferty S, De Smedt SC, Weiss R, Sanders NN, Kitada T. N1-methylpseudouridine-incorporated mRNA outperforms pseudouridine-incorporated mRNA by providing enhanced protein expression and reduced immunogenicity in mammalian cell lines and mice. Journal of Controlled Release. 2015;**217**: 337-344

[39] Pardi N, Hogan MJ, Naradikian MS, Parkhouse K, Cain DW, Jones L, et al. Nucleoside-modified mRNA vaccines induce potent T follicular helper and germinal center B cell responses. Journal of Experimental Medicine. 2018;**2156**:1571-1588

[40] Chatterjee S, Pal JK. Role of 5′-and 3′-untranslated regions of mRNAs in human diseases. Biology of the Cell. 2009;**1015**:251-262

[41] Lundstrom K. Latest development on RNA-based drugs and vaccines. Future Science OA. 2018;**45**:FSO300

[42] Mauro VP, Chappell SA. A critical analysis of codon optimization in human therapeutics. Trends in Molecular Medicine. 2014;**2011**:604-613

[43] Fåhraeus R, Marin M, Olivares-Illana V. Whisper mutations: Cryptic messages within the genetic code. Oncogene. 2016;**3529**:3753-3759

[44] Diebold SS, Kaisho T, Hemmi H, Akira S, e Sousa C R. Innate antiviral responses by means of TLR7-mediated recognition of single-stranded RNA. Science. 2004;**3035663**:1529-1531

[45] Pichlmair A, Schulz O, Tan C-P, Rehwinkel J, Kato H, Takeuchi O, et al. Activation of MDA5 requires higher-order RNA structures generated during virus infection. Journal of Virology. 2009;**8320**:10761-10769

[46] Iavarone C, O'hagan DT, Yu D, Delahaye NF, Ulmer JB. Mechanism of action of mRNA-based vaccines. Expert Review of Vaccines. 2017;**169**:871-881

[47] De Beuckelaer A, Pollard C, Van Lint S, Roose K, Van Hoecke L, Naessens T, et al. Type I interferons interfere with the capacity of mRNA lipoplex vaccines to elicit cytolytic T cell responses. Molecular Therapy. 2016; **2411**:2012-2020

[48] Kariko K, Muramatsu H, Ludwig J, Weissman D. Generating the optimal mRNA for therapy: HPLC purification eliminates immune activation and improves translation of nucleoside-modified, protein-encoding mRNA. Nucleic Acids Research. 2011; **3921**:e142-e142

[49] Blakney AK, Ip S, Geall AJ. An update on self-amplifying mRNA vaccine development. Vaccine. 2021;**92**:97

[50] Minnaert A-K, Vanluchene H, Verbeke R, Lentacker I, De Smedt SC, Raemdonck K, et al. Strategies for controlling the innate immune activity of conventional and self-amplifying mRNA therapeutics: getting the message across. Advanced Drug Delivery Reviews. 2021;**176**:113900

[51] Beissert T, Koste L, Perkovic M, Walzer KC, Erbar S, Selmi A, et al.

Improvement of in vivo expression of genes delivered by self-amplifying RNA using vaccinia virus immune evasion proteins. Human Gene Therapy. 2017;**2812**:1138-1146

[52] Beissert T, Perkovic M, Vogel A, Erbar S, Walzer KC, Hempel T, et al. A trans-amplifying RNA vaccine strategy for induction of potent protective immunity. Molecular Therapy. 2020; **281**:119-128

[53] Samnuan K, Blakney AK, McKay PF, Shattock RJ. Design-of-Experiments In Vitro Transcription Yield Optimization of Self-Amplifying RNA. bioRxiv. 2021:1-38

[54] Lundstrom K. Alphavirus-based vaccines. Viruses. 2014;**66**:2392-2415

[55] Mogler MA, Kamrud KI. RNA-based viral vectors. Expert Review of Vaccines. 2015;**142**:283-312

[56] Tonkin DR, Whitmore A, Johnston RE, Barro M. Infected dendritic cells are sufficient to mediate the adjuvant activity generated by Venezuelan equine encephalitis virus replicon particles. Vaccine. 2012;**3030**:4532-4542

[57] Krieg PA, Melton D. Functional messenger RNAs are produced by SP6 in vitro transcription of cloned cDNAs. Nucleic Acids Research. 1984;**1218**:7057-7070

[58] Krieg PA, Melton D. [25] In vitro RNA synthesis with SP6 RNA polymerase. Methods in Enzymology. 1987;**155**:397-415

[59] Pascolo S. Messenger RNA-based vaccines. Expert Opinion on Biological Therapy. 2004;**48**:1285-1294

[60] Kim J, Eygeris Y, Gupta M, Sahay G. Self-assembled mRNA vaccines. Advanced Drug Delivery Reviews. 2021;**170**:83-112

[61] Buschmann MD, Carrasco MJ, Alishetty S, Paige M, Alameh MG, Weissman D. Nanomaterial delivery systems for mRNA vaccines. Vaccine. 2021;**91**:65

[62] Ho W, Gao M, Li F, Li Z, Zhang XQ, Xu X. Next-generation vaccines: Nanoparticle-mediated DNA and mRNA delivery. Advanced Healthcare Materials. 2021;**108**:2001812

[63] Wu MZ, Asahara H, Tzertzinis G, Roy B. Synthesis of low immunogenicity RNA with high-temperature in vitro transcription. RNA. 2020;**263**:345-360

[64] Gholamalipour Y, Johnson WC, Martin CT. Efficient inhibition of RNA self-primed extension by addition of competing 3′-capture DNA-improved RNA synthesis by T7 RNA polymerase. Nucleic Acids Research. 2019;**4719**: e118-e118

Chapter 4

Maximizing COVID-19 Vaccine Acceptance in Developing Countries

Yusuff Tunde Gbonjubola, Daha Garba Muhammad,
Nwaezuoke Chisom Anastasia and Tobi Elisha Adekolurejo

Abstract

Coronavirus disease 2019 (COVID-19) is still in existence, with the capacity to spread even further. Vaccination could efficiently reduce the burden of the pandemic, but first, people must accept these vaccines. Vaccine acceptance by the population is crucial to control the pandemic and prevent further deaths. Herd Immunity, which is the indirect protection that occurs when a sufficient percentage of a population has become immune to an infection, offers some protection to unvaccinated individuals. However, herd immunity is compromised when widespread vaccine acceptance is not achieved. Some vaccines have been authorized to prevent COVID-19, such as Pfizer-BioNTech, Moderna, Johnson & Johnson's Janssen, and Oxford-AstraZeneca COVID-19 Vaccine. While vaccine development has been achieved within a short time, its safety, potency, efficacy, and universal accessibility are of great concern and could influence vaccine acceptance. Conspiracy beliefs rampant in Africa may influence vaccine hesitance; exposure to anti-vaccine theories decreases willingness to accept vaccination. As such, there is a need for the availability of reliable information about vaccines, messages that highlight the vaccines efficacy and safety could be effective for addressing the hesitancy to increase the acceptance level of the COVID-19 Vaccine in Africa.

Keywords: COVID-19, vaccine, pandemic, herd immunity

1. Introduction

Coronavirus disease 2019 (COVID-19) is still in existence. It is caused by the severe acute respiratory syndrome coronavirus 2 (SARS-CoV-2), which started in Wuhan in China [1]. While the third wave of COVID-19 in Africa continues to wane, 108,000 new cases were reported and more than 3000 people died in the week leading up to September 19. The continent currently has nearly 8.2 million cases of COVID-19. The Delta variant has been found in 38 African countries. Alpha has been found in 45 countries, and beta has been found in 40 [2]. According to the European Centre for Disease Prevention and Control (ECDPC), all continents have reported confirmed cases of the virus [3]. COVID-19 is transmitted through respiratory droplets [4]. Transmission rates are reported to be unknown for SARS-CoV-2; however, many authors have reported transmission by direct contact with an infected person. The mode of transmission of SARS-CoV-2 is as other pulmonary diseases, such as influenza [5]. Persons can be infected through contact with a contaminated surface. The virus remains viable on surfaces for lengthy hours. Van Doremalen et al. in a study reported that SARS-CoV-2 remained viable on surfaces for up to 72 hours [6]. Studies have reported that people with COVID-19 who are asymptomatic are still contagious; this has brought about the question: what is the effectiveness of isolation? [7, 8]. Zou et al. [9] reported an increased viral load in symptomatic persons while an asymptomatic person was shedding the virus.

Clinical signs of an infection include fever, headache, dry cough, shortness of breath, and fatigue [10, 11]. Some patients also report digestive symptoms such as diarrhea and vomiting [12]. COVID-19 was observed to have a similar clinical manifestation as SARS [13]. Fever, for example, occurred in 98–100% of patients with SARS compared with 81.3% of patients with COVID-19 [14, 15]. In total, 18.7% of patients had no fever at admission; this means the absence of fever does not necessarily rule out the possibility of COVID-19 [13].

Deaths and disability were directly linked with the first wave of COVID-19, alongside some population living with the aftermath of a severe acute respiratory syndrome, which could persist even after they are clinically cured of the infection [16]. Second-wave victims were those that suffer from the consequence of the measures taken to limit the spread of COVID-19. The victims include but not limited to those who did not present at the hospital due to fear of getting infected; some group of people with progressive disease whose appointments were rescheduled, and those that did not present themselves for routine screening [17]. The third wave was found to be more dangerous [18]. The rate of infection was 1.7 times more than that of second wave and 6.23 times more than the first wave. Although, the death rate was 1.21 times more than second wave, third wave was not as fatal as the first wave (at least 0.46 times less than the first wave) [19]. The third wave has been associated with the effect of the pandemic on the social determinants of health and its effect on the next generation [20]. The health inequalities have been projected to worsen through severe economic set back [21], and the groups at the intersection between poverty and poor health with most likely suffer the most [21].

Prevention of COVID-19 infection is based on adherence to social distancing, patient isolation, quarantine, wearing a mask, and regular washing of hands [22]. In addition to the health impact of COVID-19, there is an economic consequence of this virus as it spans through an increase in unemployment rates and healthcare demand [23]. These negative impacts have encouraged pharmaceutical companies to develop a vaccine urgently. In December 2020, several vaccines were authorized to prevent COVID-19 infection, and more than 50 COVID-19 vaccine candidates are being developed [24]. Vaccination is the most effective means of handling infectious diseases, and its success is confronted by vaccine hesitance [25]. The impact of vaccination against diseases cannot be overemphasized; for example, smallpox was eradicated by vaccine administration [25].

2. COVID-19 vaccine development

Several vaccines for COVID-19 have undergone human trials and passed. Three vaccines have been approved for administration, and they are Pfizer-BioNTech, Moderna, and Johnson & Johnson's Janssen. The Centers for Disease Control and Prevention (CDC) has declared that all three vaccines are safe, effective, and will reduce a person's risk of severe illness [26]. The Bill and Melinda Gates Foundation awarded a US$15 million grant toward the experimental COVID-19 Vaccine by Novavax with research stationed at Witwatersrand University in Johannesburg, South Africa [27]. While vaccine development has been achieved within a short time, its safety, potency, efficacy, and universal accessibility are of great concern [12]. Efficacy is not all required for vaccines to be effective, but it must be accepted by the people [28].

3. Mechanism of action of COVID-19 vaccines

3.1 mRNA vaccines

3.1.1 Pfizer and Moderna

Both Pfizer and Moderna are mRNA vaccines. These are cutting-edge vaccines made from genetically engineered RNA molecules that through their translation in the ribosomes produce a protein that triggers the immune system of the host to produce antibodies against the antigen-carrying substance or organism [29]. The surface of COVID 19 contains between 25 and 28 proteins, and on this surface is the presence of three repeating protein copies called "spike protein" of COVID 19, but only a copy of one of this repeating subunit is used for formulating the mRNA [30]. This formulated mRNA is enclosed in a lipid nanoparticle to ensure cytoplasmic penetration and ribosome translation leading to the synthesis of the viral spike proteins [29] as well as making delivery easy and preventing the body from damaging it [31]. The synthesized s-proteins displayed on cell membranes activate MHC 1 and MHC 2 complexes and activate the B cells, macrophages, dendtritic cells and attract other cells of the immune system such as T helper cells [32]. Strongly activated T helper cells produce interleukins 2, 4, and 5 [33]. These interleukins stimulate the T helper cells to proliferate memory T cells for recognition of the spike protein and also cause the B-cells to differentiate into plasma cells that eventually produce antibodies against the viral spike proteins, neutralizing or destroying the virus. The efficacy of Pfizer and Moderna for preventing disease or severe disease results in 95–87.5% and 94.5–100%, respectively [32].

3.2 Viral vector vaccines

3.2.1 Astra Zeneca

Astra Zeneca is a viral vector vaccine. This involved the use of a modified version of chimpanzee DNA adenovirus known as ChAdOx1, which no human population has been exposed and does not generate an immune response to the adenovirus but rather to the viral proteins encoded in the host DNA [34]. The genetically engineered DNA vector is used as a template in human cells to generate new chimpanzee adenovirus replicas and produce the viral proteins that is similar to the s-peptide and therefore elicits an immune response. The DNA vector on entry into the cytoplasm of host cells migrates to the cell nucleus, where it gets converted into mRNA using the host enzymes. The mRNA migrates back to the cytoplasm and interacts with the ribosome to synthesize the s-proteins. The proteins get expressed on the cell membranes and form MHC 1and MHC 2 complexes. This is followed by the activation of antibodies, B cells, T helper cells, and plasma cells against spike proteins of COVID 19, destroying or neutralizing the virus [35].

3.2.2 Johnson and Johnson

Johnson and Johnson vaccine, also known as Janssen COVID 19 vaccine, is also a viral vector vaccine made from an adenovirus type 26 and genetically engineered to contain the gene for making spiked proteins of SARS-CoV-2 (glycoprotein (Ad26. COV2-S). The mechanism of action is similar to that of Astra Zeneca. The adenovirus vector is manipulated in the laboratory to delete the gene for replication to avoid replication in human cells [36]. When injected into human body, the DNA carrying the information for the SARS-CoV-2 spike protein is transferred into the nucleus without being incorporated into the host cell DNA. The viral DNA causes the cell to produce more adenovirus particles. It gets translated into mRNA and transported out to the cytoplasm. Ribosomes convert information in these mRNA to form spike proteins. These spike proteins present at the surface of cells induce production of T-cells (CD4 and CD8), cytotoxic cells, plasma cells, Interleukins, and B-cells that constitute the three primary immune responses (antibodies, killer CD8 T-cells, and helper CD4T-cells) to block the infection [37]. T cells destroy infected human cells while antibodies protect uninfected cells from circulating free viral particles.

The efficacy of Astra-Zeneca and Janssen is about 70% and 65%, respectively; in the case of Janssen, it depends on the geographical area ranging from 72–57% [32].

4. COVID-19 vaccine acceptance

Vaccine acceptance by the population is crucial to control the pandemic and prevent further deaths. Herd Immunity, also known as population immunity, offers some protection to unvaccinated individuals when a significant percentage of the population is immune to that disease. Herd immunity, however, is compromised when widespread vaccine acceptance is not achieved [38]. Vaccination could efficiently reduce the burden of the pandemic [39]. However, a reasonable level of vaccine acceptance is needed [40]. As the vaccine is now available, its public acceptance is about 67% in the United States of America [41]. From the history books, Africa has always been a passive recipient of vaccines, and the reason for this is multifactorial [26]. Vaccines have been a successful measure of disease prevention for decades [42]. However, vaccine hesitancy and refusal are significant global concerns, prompting the World Health Organization (WHO) to declare this uncertainty among the top 10 health threats in 2019 [43].

5. Herd immunity and COVID-19 vaccine

Herd immunity, first published in 1923, is a concept that refers to the reduced risk of an individual getting an infection from an already infected individual because of the resistant immunity of a large proportion of the population [44]. With this concept, the fight against infection in a population can be achieved even when it is impossible to vaccinate the entire population [45]. Immunization coverage of about 80% of individuals against the smallpox virus reduced the transmission rate and eradicated the disease [46]. Individuals with compromised immunity as a result of diseases such as HIV/AIDS may not be able to be vaccinated, but they are still being protected from the infection through herd immunity, by staying among people who have been vaccinated [47]. The administration of COVID-19 vaccines is expanding daily, and this brings us a step closer to the goal of COVID-19 herd immunity [26, 48].

Herd immunity, also known as indirect protection, community immunity, or community protection, refers to the protection of susceptible individuals against an infection when a sufficiently large proportion of immune individuals exist in a population [49]. In other words, herd immunity is the inability of infected individuals to propagate an epidemic outbreak due to lack of contact with sufficient numbers of susceptible individuals [49]. It was initially introduced more than a century ago and in the latter half of the twentieth century, the expansion of immunization programs and the need for describing targets for immunization coverage, discussions on disease eradication, and cost-effectiveness analyses of vaccination programs make the term "herd immunity" more popular [50]. The disappearance of smallpox alongside the sustained reduction in the incidence of disease in elderly population who were not vaccinated following routine childhood immunization with conjugated *Haemophilus influenzae* type B and pneumococcal vaccines are successful examples of the effects of vaccine-induced herd immunity [50].

The herd immunity threshold is defined as the proportion of individuals in a population who, having acquired immunity, can no longer participate in the chain of transmission [51]. If the population of those with immunity falls beyond this threshold, current outbreaks will extinguish and endemic transmission of the pathogen will be interrupted [51]. The durability of immune memory is a critical factor in determining population-level protection and sustaining herd immunity in both naturally acquired and vaccine-induced immunity [52]. In the case of measles, varicella, and rubella, long-term immunity has been achieved with both infection and vaccination. But such durable immunity has not been observed in coronal virus [52].

Herd immunity is an important defense against outbreaks and has shown success in regions with satisfactory vaccination rates [49]. Notable is that even small deviations from protective levels can allow for significant outbreaks due to local clusters of susceptible individuals, as has been seen with measles over the past few years. As such Saad [49] concluded that vaccines must not only be effective, but vaccination programs must be efficient and broadly adopted to ensure that those who cannot be directly protected will nonetheless derive relative protections.

6. Hesitancy toward the COVID-19 vaccine

Vaccine hesitancy is defined as a delay in the acceptance or refusal of a vaccine despite the availability of vaccine services [53]. Vaccine hesitancy is a public health threat as it serves as a barrier to immunization coverage, especially in developing countries [46]. Conspiracy beliefs, which are common in Africa, may influence vaccine hesitance as exposure to anti-vaccine theories decreases willingness to accept vaccination [43]. Different studies across the world have shown varying rates of vaccine acceptance among people [41, 54–57]. The results of these studies show that there is some vaccine hesitancy toward the COVID-19 vaccines among different demographics. In a study conducted among 672 adults in the United States, 67% said that they would accept a COVID-19 vaccine if it is recommended for them [41]. Another study conducted across 19 countries reported a 71.5% acceptance rate among participants [56]. In Africa, the hesitancy toward the vaccines appears to be larger.

According to the latest data from South African fintech company, Comparisure, only 48% of South Africans said they would take a COVID-19 vaccine if it was available [21]. In a study conducted among Cameroonians, the vaccine hesitancy rate was 84.6%, and this percentage includes participants who said they will need more information, do not know or will not take a COVID-19 vaccine [58].

7. Factors that cause hesitancy toward COVID-19 vaccines

The causes of vaccine hesitancy toward the COVID-19 vaccines are intersectional. One of the factors that affect the vaccine-acceptance rate among people is the efficacy of the vaccine. In a study conducted in South-East Asia, 93.3% of respondents said they would like to be vaccinated for a 95% effective vaccine, but the acceptance rate dropped to 67% for a vaccine with 50% effectiveness [55]. Other factors that influence vaccine hesitancy include a perception that vaccines are not beneficial, pain and needle fear, negative information about vaccination on social media, and a lack of knowledge about vaccines [58]. Lower educational attainment and a lower household income are factors that can drive vaccine hesitancy [55]. The causes of vaccine hesitancy had also been reported in different studies, including religious reasons, personal beliefs, and safety concerns due to widespread myths, including the association of vaccines and autism, brain damage, and other conditions [59].

8. Consequences of vaccine hesitancy toward COVID-19 vaccines

Vaccine hesitancy has many consequences. These consequences can be public-health-related and even socioeconomic. Vaccine hesitancy will increase the incidence of COVID-19 infections. Herd Immunity, which offers some protection to unvaccinated individuals, is compromised when widespread vaccine acceptance is not achieved [38]. One study found that a 5% reduction in measles, mumps, and rubella vaccination in the United States resulted in a threefold increase in annual measles cases [60].

9. Approach to maximizing COVID-19 vaccine acceptance

Community participation has been identified as a key approach to maximizing COVID-19 acceptance in developing countries. Mobilization of community members can be achieved through religious leaders, chiefs, and royal heads' involvement in the process [61–63]. There are two major goals of community mobilization. The first is to ensure proper education on the benefits of the vaccine with the assistance of health professionals [61–63]. Second is to maximize uptake of COVID-19 vaccine when it becomes available to the public. Failure of the community to accept the vaccine would result in adverse consequences such as resource wastage. Community participation will make enhance and promote the availability of the vaccine in each African setting through a well-structured planning process of vaccine production and procurement [61–63]. More so, integration of the COVID-19 vaccine into the existing healthcare services presents a promising strategy to overcome the problem of vaccine hesitancy to improve vaccine acceptance maximization [61–63]. Integration of the COVID-19 vaccine in healthcare facilities reduces the time people spend on vaccine collection. It is worthy of note that the integration of the COVID-19 vaccine in health facilities (primary healthcare centers) promotes its proximity to residential areas, thus reducing the individual's cost of transportation [61–63]. The National Primary Health Care Development Agency should be both responsive and responsible in this regard.

10. Conclusion

Vaccine hesitancy may hinder herd immunity, an important aspect of curtailing COVID-19. Efficacy of the vaccine, misinformation on the vaccine by religious leaders and the masses are part of the causes of hesitancy. Acceptance of the COVID-19 vaccine by the general population is so important for achieving a wide range of immunization coverage to bring the pandemic to an end. Community mobilization by community elders has been suggested as a way of getting to people in the community and could improve vaccine acceptance in Africa. There is also a need for the availability of reliable information about vaccines, messages that highlight the vaccines efficacy and safety could be effective for addressing the hesitancy to increase the acceptance level of the COVID-19 vaccine in Africa.

Conflict of interest

Nil

Author details

Yusuff Tunde Gbonjubola[1], Daha Garba Muhammad[2*],
Nwaezuoke Chisom Anastasia[3] and Tobi Elisha Adekolurejo[4]

1 Health Services Department, Abubakar Tafawa Balewa University, Bauchi, Nigeria

2 Dutse General Hospital, Jigawa State, Nigeria

3 Federal Medical Centre Abuja, Nigeria

4 Department of Physiotherapy, University of Ibadan, Nigeria

*Address all correspondence to: dahagarba@gmail.com

IntechOpen

References

[1] Zhu NA et al. A novel coronavirus from patients with pneumonia in China, 2019. New England Journal of Medicine. 2020;**382**(8):727-733

[2] European Centre for Disease Prevention and Control. 2020. COVID-19 situation update worldwide, as of August 31 2020. Retrieved March 30, 2021 from: https://www.ecdc. europa.eu/en/geographical-distribution-2019-ncov-cases

[3] Sun P, Lu X, Xu C, Sun W, Pan B. Understanding of COVID-19 based on current evidence. Journal of Medical Virology. 2020;**92**(6):548-551

[4] Center for Disease Control and Prevention. 2020. Retrieved March 31, 2021 from: https://www.cdc.gov/ coronavirus/2019-ncov/vaccines/ different-vaccines.html

[5] Van Doremalen N, Bushmaker T, Morris DH, Holbrook MG, Gamble A, Williamson BN, et al. Aerosol and surface stability of SARS-CoV-2 as compared with SARS-CoV-1. The New England Journal of Medicine. 2020; **382**(16):1564-1567

[6] Bai Y, Yao L, Wei T, Tian F, Jin DY, Chen L, et al. Presumed asymptomatic carrier transmission of COVID-19. JAMA. 2020;**323**(14):1406-1407

[7] Yu P, Zhu J, Zhang Z, Han Y, Huang L. A familial cluster of infection associated with the 2019 novel coronavirus indicating potential person-to-person transmission during the incubation period. The Journal of Infectious Diseases. 2020;**221**(11):1757-1761. DOI: 10.1093/infdis/jiaa077

[8] Zou L, Ruan F, Huang M, Liang L, Huang H, Hong Z, et al. SARS-CoV-2 viral load in upper respiratory specimens of infected patients. The New England Journal of Medicine.

2020;**382**(12):1177-1179. DOI: 10.1056/ NEJMc2001737. Epub 2020 February 19

[9] Huang C, Wang Y, Li X, Ren L, Zhao J, Hu Y, et al. Clinical features of patients infected with 2019 novel coronavirus in Wuhan, China. The Lancet. 2020;**395**(10223):497-506

[10] Lai CC, Liu YH, Wang CY, Wang YH, Hsueh SC, Yen MY, et al. Asymptomatic carrier state, acute respiratory disease, and pneumonia due to severe acute respiratory syndrome coronavirus 2 (SARSCoV-2): Facts and myths. Journal of Microbiology, Immunology and Infection. 2020; **53**(3):404-412

[11] Wang J, Peng Y, Xu H, Cui Z, Williams RO 3rd. The COVID19 vaccine race: Challenges and opportunities in vaccine formulation. AAPS PharmSciTech. 2020;**21**(6):225

[12] Mo P, Xing Y, Xiao Y, et al. Clinical characteristics of refractory COVID-19 pneumonia in Wuhan, China. Clinical Infectious Diseases. 2020;**73**(11): e4208-e4213. DOI: 10.1093/cid/ciaa270

[13] Assiri JA, Al-Tawfiq, Al-Rabeeah AA, et al. Epidemiological, demographic, and clinical characteristics of 47 cases of Middle East respiratory syndrome coronavirus disease from Saudi Arabia: A descriptive study. The Lancet Infectious Diseases. 2013;**13**(9):752-761

[14] Yin Y, Wunderink RG. MERS, SARS and other coronaviruses as causes of pneumonia. Respirology. 2018; **23**(2):130-137

[15] Callard F. Very, very mild: covid-19 symptoms and illness classification [Internet]. Somatosphere. 2020. Available from: somatosphere.net/2020/ mild-covid.html/ [Accessed: 31 May 2020]

[16] Fisayo T, Tsukagoshi S. Three waves of the COVID-19 pandemic. Postgraduate Medical Journal. 2020;**97**:332. DOI: 10.1136/postgradmedj-2020-138564.2021

[17] Seong H, Hyun HJ, Yun JG, Noh JY, Cheong HJ, Kim WJ, et al. Comparison of the second and third waves of the COVID-19 pandemic in South Korea: Importance of early public health intervention. International Journal of Infectious Diseases. 2021;**104**:742-745

[18] Basak A, Rahaman S, Guha A. Dynamics of the third wave, modelling COVID-19 pandemic with an outlook towards India; 2021. DOI: 10.1101/2021.08.17.21262193

[19] Major LE, Machin S. Covid-19 and Social Mobility [Internet]. London, UK: London School of Economics and Political Science; 2020. Available from: http://cep.lse.ac.uk/pubs/download/cepcovid-19-004.pdf

[20] Banks J, Karjalainen H, Propper C. Recessions and Health: The Long-Term Health Consequences of Responses to the Coronavirus [Internet]. London, UK: Institute for Fiscal Studies; 2020. Available from: https://www.ifs.org.uk/publications/14799

[21] WHO. 2021. Weekly epidemiological report. Retrieved on April 18, 2021 from: https://www.who.int/en/news-room/fact-sheets/detail/Weekly-epidemiological-report

[22] Abdi M. Coronavirus disease 2019 (COVID-19) outbreak in Iran; actions and problems. Infection Control and Hospital Epidemiology. 2020;**41**(6):1-5

[23] Ita K. Coronavirus disease (COVID-19): Current status and prospects for drug and vaccine development. Archives of Medical Research. 2021;**52**(1):15-24

[24] KabambaNzaji M, Kabamba Ngombe L, NgoieMwamba G, BanzaNdala DB, MbidiMiema J, LuhataLungoyo C, et al. Acceptability of vaccination against COVID-19 among healthcare Workers in the Democratic Republic of the Congo. Pragmatic and Observational Research. 2020;**11**:103-109

[25] Lucero-Prisno DE, Adebisi YA, Uzairue LI, et al. Ensuring access to COVID-19 vaccines among marginalised populations in Africa. Public Health. 2020;**197**:e14-e15. DOI: 10.1016/J.puhe.2022.01.008

[26] Center for Disease Control and Prevention. 2021. Retrieved March 31, 2021 from: https://www.cdc.gov/coronavirus/2019

[27] Makoni M. COVID-19 vaccine trials in Africa. The Lancet Respiratory Medicine. 2020;**8**(11):e79-e80

[28] Callaway E. Outrage over Russia's fast-track coronavirus vaccine. Nature. 2020;**584**(7821):334-335

[29] Sundaram N, Schaetti C, Merten S, Schindler C, Ali SM, Nyambedha EO, et al. Sociocultural determinants of anticipated oral cholera vaccine acceptance in three African settings: A meta-analytic approach. BMC Public Health. 2016;**16**:36

[30] Zhou X, Jiang X, Qu M, et al. Engineering antiviral vaccines ACS nano. ACS Nano. 2020;**1**(14):12370-12389

[31] Chen Y, Liu Q, GuO D. Emerging coronavirus: Genome structure, replication, and pathogenesis. Journal of Medical Virology. 2020;**92**(4):418-423. DOI: 10.1002/jmu.25681

[32] Centers for Disease Control and Prevention. COVID-19. mRNA Vaccines. Availabe from: https:www.cdc.gov/vaccines/covid-19/clinical-consideration/covid-19-vaccines-us.html. [Accessed Nov. 3, 2021]

[33] Mascellino MT, Di Timoteo F, De Angelis M, Oliva A. Overview of the main anti-SARS-CoV 2 vaccines: Mechanism of action, efficacy and safety. Journal abbreviation. Infection and Drug Resistance. 2021;**31**(14):3459-3476. DOI: 10.2147/IDR.S315727

[34] Cox RJ, Brokstad KA. Not just antibodies: b cells and T cells mediate immunity to COVID-19. Nature Reviews. Immunology. 2020;**20**(10):581-582. DOI: 10.1038/s41577-020-00436-4

[35] Ninja Nerd Lectures. Covid-19 Vaccines: Moderna and Pfizer-BioNTech. 2020. Available from: http://wwwyoutube.com/watch?v=35ldbICU4o [Accessed: July 29, 2021]

[36] Sette A, Crotty S. Adaptive immunity to SARS-CoV-2 and COVID-19. Cell. 2021;**184**(4):861-880. DOI: 10.1016/j.cell.2021.01.007

[37] Covid-19 Info Vaccine. Retrieved From: https://www.covid19infovaccines.com/posts/how-does-the-oxford-astrazeneca-covid-19-vaccine-work [Accessed: December 13, 2021]

[38] Grifoni A, Weiskopf D, Ramirez S, et al. Targets of T cell responses to SARS-CoV-2 coronavirus in humans with COVID-19 disease and unexposed individuals. Cell. 2020;**181**(7):1489-1501

[39] Dzinamarira T, Nachipo B, Phiri B, Musuka G. COVID-19 vaccine roll-out in South Africa and Zimbabwe: Urgent need to address community preparedness, fears and hesitancy. Vaccine. 2021;**9**(3):250. DOI: 10.3390/vaccines9030250

[40] Walsh EE, Frenck RW, Falsey AR, Kitchin N, Absalon J, Gurtman A. Safety and immunogenicity of two RNA-based Covid-19 vaccine candidates. The New England Journal of Medicine. 2020;**383**:2439-2450

[41] Ditekemena JD, Nkamba DM, Mutwadi A, Mavoko HM, SieweFodjo JN, Luhata C, et al. COVID-19 vaccine acceptance in the Democratic Republic of Congo: A cross-sectional survey. Vaccine. 2021;**9**(2):153

[42] Zhu NA et al. A novel coronavirus from patients with pneumonia in China, 2019. New England Journal of Medicine. 2020;**382**(8):727-733. (Basel) 9(2):153

[43] Malik AA, McFadden SA, Elharake J, Omer SB. Determinants of COVID-19 vaccine acceptance in the US. EClinicalMedicine. 2020;**26**:100495

[44] Vaccines and Immunization. Available online at: http://www.who.int/healthtopics/vaccine-and-immunization?

[45] Geoghegan S, O'Callaghan KP, Offit PA. Vaccine safety: Myths and misinformation. Frontiers in Microbiology. 2020;**11**:372

[46] Goncalves G. Herd immunity: Recent uses in vaccine assessment. Expert Review of Vaccines. 2008;**7**:1493-1506

[47] Rashid H, Khandaker G, Booy R. Vaccination and herd immunity: What more do we know? Current Opinion in Infectious Diseases. 2012;**25**(3):243-249

[48] WHO. Immunization Coverage. 2019. Retrieved on April 18, 2021 from: https://www. who.int/en/news-room/fact-sheets/detail/immunization-coverage

[49] World Health Organization. How do vaccines work? 2020. Retrieved August 8, 2021 from: https://www.who.int/news-room/feature-stories/detail/how-do-vaccines-work?gclid=CjwKCAjw9aiIBhA1EiwAJ_GTSmjiXu30XBj6XERluShYPRbo6ZTkbpiT7AVKwSVpPPwkjD702GvRpxoCxfUQAvD_BwE

[50] Frederiksen LSF, Zhang Y, Foged C, Thakur A. The long road toward

COVID-19 herd immunity: Vaccine platform technologies and mass immunization strategies. Frontiers in Immunology. 2020;**11**:1817

[51] Saad B, Yildirim I, Howard P. Herd immunity and implications for SARS-CoV-2 control. JAMA. 2020, 2020; **324**(20):2095-2096. DOI: 10.1001/jama.2020.20892

[52] Fine P, Eames K, Heymann DL. "Herd immunity": A rough guide. Clinical Infectious Diseases. 2011;**52**(7):911-916. DOI: 10.1093/cid/cir007PubMedGoogle ScholarCrossref

[53] Van den Driessche P. Reproduction numbers of infectious disease models. Infectious Disease Modelling. 2017;**2**(3):288-303

[54] Edridge AWD, Kaczorowska J, ACR H, et al. Seasonal coronavirus protective immunity is short-lasting. Nature Medicine. 2020;**33**(34):4161-4164. DOI: 10.1038/s41591-020-1083-1

[55] MacDonald, N. E., & SAGE Working Group on Vaccine Hesitancy. Vaccine hesitancy: Definition, scope and determinants. Vaccine. 2015; **33**(34):4161-4164. DOI: 10.1016/j.vaccine.2015.04.036

[56] Fisher KA et al. Attitudes toward a potential SARS-CoV-2 vaccine. Annals of Internal Medicine. 2020;**173**(12):964-973. DOI: 10.7326/m20-3569

[57] Harapan H, Wagner AL, Yufika A, Winardi W, Anwar S, Gan AK, et al. Acceptance of a COVID-19 vaccine in Southeast Asia: A cross-sectional study in Indonesia. Frontiers in Public Health. 2020;**14**(8):381. DOI: 10.3389/fpubh.2020.00381

[58] Lazarus JV, Ratzan SC, Palayew A, et al. A global survey of potential acceptance of a COVID-19 vaccine. Nature Medicine. 2021;**27**:225-228. DOI: 10.1038/s41591-020-1124-9

[59] Jolley D, Douglas KM. The effects of anti-vaccine conspiracy theories on vaccination intentions. PLoS One. 2014;**9**(2):e89177

[60] Dinga JN, Sinda LK, Titanji VPK. Assessment of vaccine hesitancy to a COVID-19 vaccine in Cameroonian adults and its global implication. Vaccine. 2021;**9**(2):175. DOI: 10.3390/vaccines9020175

[61] McKee C, Bohannon K. Exploring the reasons behind parental refusal of vaccines. The Journal of Pediatric Pharmacology and Therapeutics. 2018;**21**:104-109

[62] Lo NC, Hotez PJ. Public health and economic consequences of vaccine hesitancy for measles in the United States. JAMA Pediatrics. 2017; **171**(9):887-892. DOI: 10.1001/jamapediatrics.2017.1695

[63] Ilesanmi O, Afolabi A. In search of the true prevalence of COVID-19 in Africa: Time to involve more stakeholders. International Journal of Health and Life Sciences. 2020; In Press (In Press):e108105

Chapter 5

Helicobacter pylori Challenge Vaccine for Humans

Rike Syahniar, Dayu Swasti Kharisma and Rayhana

Abstract

Helicobacter pylori infect during childhood and are typically present for life, despite a vigorous host immune response, which includes the invading pathogen being coated with antibodies. This bacterial longevity indicates the development, on the part of the pathogen, of a range of processes for evading effective host immunity. Since its discovery 25 years ago, significant progress has been made in understanding the virulence factors and several aspects of the pathogenesis of *H. pylori* gastric diseases. The prevalence of antimicrobial drug resistance is so high that all patients infected with *H. pylori* should be considered resistant infections. The most severe consequence of *H. pylori* infection, and the key reason a vaccine is required, is gastric cancer, globally the third leading cause of death due to cancer. Patients typically present with gastric cancer without knowing they are infected; eradication likely has little effect by this time. Vaccine against *H. pylori* that reduces the incidence of gastric cancer will probably be cost effective in developed countries. Several vaccines were successfully tested in different experimental animal models, but translation into an efficacious human vaccine has been unsuccessful.

Keywords: *Helicobacter pylori*, *H. pylori* vaccine, gastric disease, gastric cancer

1. Introduction

Helicobacter pylori (H. pylori) is a gram-negative, motile, spiral, or curved bacterium that colonizes the human gastric mucosa about 50% of the human population [1]. *H. pylori* induces the development of a peptic ulcer, gastric mucosa-associated lymphoid tissue lymphoma, and gastric cancer [2]. Globally, gastric cancer ranks as the third leading cause of death from malignancy [3].

The prevalence of *H. pylori* infection in developed countries ranges from 30 to 40%, while it can reach 80% [4–8]. Hygiene conditions and socioeconomic status affect the incidence of *H. pylori* infection [5]. *H. pylori* infection is thought to result from direct human-to-human transmission through oral, fecal, or both. The infection is generally acquired during childhood and then increases gradually with age. *H. pylori* is commonly transmitted from mother to child [1] This evidence is supported by the ability to grow *H. pylori* from vomit or oral area and analysis of bacterial strains indicating general vertical transmission between mothers and their offspring [1, 9, 10]. If left untreated, most *H. pylori* infections last a lifetime.

It is challenging to eradicate *H. pylori* because of its high antimicrobial resistance. In addition, most of the infected people are asymptomatic. The costs for adequate diagnostic tests and pharmacological eradication will be enormous. Treatment of *H. pylori* requires a multidrug regimen because the organism colonizes beneath the mucous layer, which reduces the direct effect of antibiotic penetration [11]. Resistance is also a problem with some commonly used antibiotics, namely metronidazole, amoxicillin, erythromycin, and clarithromycin [11–15]. Unfortunately, increasing antibiotic resistance is beginning to affect the efficacy of treatment, and, despite the impact of *H. pylori*, preventive vaccination strategies are still lacking. Until now, there is no recommended *H. pylori* vaccine available. Here, we provide an overview of the constraints and challenges in the manufacture of vaccines against *H. pylori*.

2. Morphology of *H. pylori*

H. pylori is a gram-negative, non-spore-forming bacterium, spiral-shaped or rods. It will turn into a coccoid form in unfavorable conditions, a form of defense resistant to conditions [16]. Several studies have shown that the coccoid form of *H. pylori* decreases morphological manifestations and remains alive and metabolizes actively, although it cannot be cultured [16, 17].

These bacteria have flagella that allow high motility and are microaerophilic. *H. pylori* has 1–6 sheathed flagella at the terminals, which are lophotrically arranged. Other Helicobacter species have flagella that are not sheathed [18]. The length of this bacterium is between 2.5 and 3.5 m, and the width is 0.5–1.0 m. *H. pylori* can grow well at 35–37°C and produces the enzyme catalase, cytochrome oxidase, urease, alkaline phosphatase, and glutamyl transpeptidase. The proper place for bacteria to live in the human body is the antrum. The number of *H. pylori* that appears to show the ability to adapt to certain areas, for example, in the human stomach on the surface of epithelial cells and the mucous layer [16, 18].

These bacteria have the same layer composition as other gram-negative bacteria: an inner membrane, periplasm with peptidoglycan, and an outer membrane. The outer membrane is composed of phospholipids and lipopolysaccharides (LPS). *H. pylori* LPS variation plays a role in population heterogeneity and allows adaptation to changes in gastric mucosal conditions. Outer membrane phospholipids contain cholesterol glucoside, which is very rare in bacteria [18].

3. *H. pylori* adaptation and colonization

After entering the stomach, *H. pylori* performs four colonizing stages, including surviving against gastric acid, moving toward gastric epithelial cells, attaching to gastric epithelial cells, and releasing toxins, causing tissue damage [19]. The enzyme urease is an essential factor for *H. pylori* colonization of the gastric mucosa. This enzyme converts urea (a product secreted by gastric cells) into ammonia and carbon dioxide. Ammonia can increase the pH of the gastric mucosa around bacterial cells. Therefore, this enzyme neutralizes the acidity of the stomach and aids in colonization. However, this enzyme can also stimulate monocytes, neutrophil chemotaxis, and stimulate cytokine production [20].

H. pylori uses flagella and specific chemoreceptors, TlpB, to move toward gastric epithelial cells near-neutral pH. The circular motion is facilitated by the helical shape of the bacteria, which can pass through the dense mucosal layer [20]. These bacteria can penetrate the mucus layer through the production of a protein called collagenase/mucinase. Collagenase/mucinase functions to liquefy mucin, reducing viscosity and allowing these bacteria to move more freely to reach epithelial cells. In addition, *H. pylori* produces alpha-carbonic anhydrase (α-CA), which helps urease convert carbon dioxide into bicarbonate. Bicarbonate is a weak base that can neutralize stomach acidity [19].

After successfully passing gastric acid, *H. pylori* attaches to gastric epithelial cells with the help of adhesins. Adhesins bind to receptors on the gastric mucosal surface. The majority of adhesins are *H. pylori* outer membrane proteins (OMPs) [21]. These adhesives include BabA (Blood Group Antigen-Binding Adhesin), SabA (Sialic acid-binding Adhesin), AlpA/B (Adherence-associated Lipoproteins A/B), HopZ, OipA, and HpaA. Receptors for BabA, Sab A, and AlpA/B adhesins include the Lewis human blood antigen group b (Leb), sialyl Lex, and laminin. Meanwhile, the HopZ and OipA receptors have not been identified [21–23].

The LPS chemical structure of several *H. pylori* strains resembles the Lewis x and Lewis y blood antigen groups expressed in the gastric mucosa. It serves to downregulate the immune response in patients with acute and chronic infections [24]. Specific modification of the LPS molecule allows molecular mimicry and alteration of the structural components of lipid A, leading to low endotoxic activity. The *H. pylori* membrane is coated with the same molecules on the host cell as plasminogen and cholesterol that protect the bacteria from host cell recognition. The high genetic diversity of bacteria allows for rapid adaptation to environmental changes [20]. Phospholipase *H. pylori* produces products such as lysolecithin, which interfere with the protective layer of phospholipids that are abundant in the apical membrane of mucus cells [24].

H. pylori has a Cag Pathogenicity Island (CagPAI), which is associated with the development of chronic active gastritis, peptic ulcer, and atrophic gastritis with an increased risk of gastric cancer. CagA is a virulence factor located at one end of CagPAI, a 40 kb of the *H. pylori* genome. Cag PAI encodes 31 genes that make up the type IV secretory system (T4SS), which injects CagA, an oncoprotein, into the cytosol of gastric epithelial cells [23]. Upon entry into gastric epithelial cells, CagA undergoes Src-dependent tyrosine phosphorylation and activates SHP-2, which leads to dephosphorylation of host cell proteins and changes in cellular morphology. Translocation of CagA protein supports the release of essential nutrients to the apical side of epithelial cells, either by induction of inflammation or by host cell depolarization [24]. Apart from CagA, peptidoglycan is also transported into the host cell *via* T4SS and outer membrane vesicles. Peptidoglycan activates the intracellular Nod1 receptor, which activates NF-kB (a transcription factor associated with epithelial gene expression and regulates the expression of various proinflammatory cytokines). Modifications in these settings are important to dampen the host immune system and contribute to bacterial persistence [25].

The second most studied *H. pylori* virulence factor is VacA. VacA is a 140-kDa polypeptide. The gene-encoding VacA is present in all *H. pylori* strains and exhibits allelic diversity in three major regions, namely, s (signal), the i (intermediate), and m (middle). As a result, the cytotoxic activity of the toxin varies between strains. Different combinations of alleles from each region (s1, s2, i1, i2, m1, m2) that exist result in different abilities of VacA toxin to stimulate vacuolation in epithelial cells [24]. VacA can bind to several epithelial cell surface molecules, including trans-membrane protein receptors-type tyrosine-protein phosphatase (PTPRZ1), fibronectin, EGFR, CD18 on T-cells, and various lipids and sphingomyelin [24]. VacA can induce vacuolation and several cellular activities, including membrane channel formation, the release of cytochrome c from mitochondria leading to apoptosis, and binding to cell membrane receptors followed by initiation of the proinflammatory response. VacA can also inhibit the activation and proliferation of T and B cells [23]. VacA can also inhibit the phagosomal maturation of macrophages and induce macrophage apoptosis [24].

4. Natural history of *H. pylori* infection

The natural history of *H. pylori* infection can be divided into two stages. The first is the acute phase, where bacteria multiply and cause gastric inflammation, hypochlorhydria, and some gastrointestinal symptoms such as fullness in the stomach, nausea, and vomiting [26]. This phase often occurs during childhood and is almost difficult to diagnose. After several weeks, the chronic phase begins with a reduced inflammatory response, and gastric pH normalizes, then becomes asymptomatic. Colonization of *H. pylori* in the gastric mucosa causes infiltration of neutrophils and mononuclear cells in the antrum and body, leading to chronic inflammation. When colonization becomes persistent, there is a close correlation between the level of acid secretion and the distribution of chronic gastritis. The most common feature is non-atrophic gastritis with normal acid secretion in asymptomatic subjects. Antral-dominant gastritis is associated with hyperchlorhydria and duodenal ulcers, whereas dominant corpus gastritis causes hypochlorhydria, gastric atrophy, intestinal metaplasia, and an increased risk of distal gastric cancer [27].

H. pylori colonization causes a degree of inflammation in the gastric mucosa. Reactive oxygen species produced from polymorphonuclear after activation by *H. pylori* induce gastric mucosal injury. Polymorphonuclear cell infiltration of the gastric mucosa leads to the development of early *H. pylori* infection lesions called active chronic gastritis, the natural history of *H. pylori* infection [16].

Acid secretion is affected by *H. pylori* infection, which is also associated with dyspepsia. Patients with functional dyspepsia and *H. pylori* infection had a fourfold increase in acid secretion. In contrast, asymptomatic *H. pylori*-positive individuals had only a 2.5-fold increase in acid secretion. Acid secretion during *H. pylori* infection depends on the spread of gastric mucosal atrophy and the local stage of inflammation, which is determined by interactions between the host, bacteria, and environmental factors. In *H. pylori*-infected patients with dominant antral gastritis without corpus atrophy, acid secretion was more elevated than in uninfected normal mucosa. This is a potential cause of dyspeptic symptoms, such as epigastric pain or burning. In contrast, when the atrophy extends to the corpus mucosa (fundic glands), reduced acid secretion is due to direct damage to parietal cells in the corpus, associated with gastric ulcers and gastric cancer [16, 28, 29].

5. Immune evasion in persistent infection with *H. pylori*

The primary defense barrier against *H. pylori* is the mucus secreted by epithelial cells and innate immune cells in the lamina propria [30]. *H. pylori* and its products can directly contact lamina propria immune cells, resulting in an influx of immune cells that include neutrophils, macrophages, dendritic cells, lymphocytes, and associated innate and adaptive immune responses [31]. Toll-like receptors (TLRs) are a major group of Pattern Recognition Receptors (PRRs) that recognize pathogen-associated molecular patterns (PAMPs). Bacterial lipopolysaccharide (LPS), peptidoglycan, lipoprotein, lipoteichoic acid, and unmethylated CpG-rich regions of DNA are the main targets of TLRs [31]. Studies have shown that *H. pylori* has managed to escape the introduction of TLRs. For example, TLR4 runs the well-described LPS recognition [32].

H. pylori subverts the adaptive immune response by modulating effector T cells [31]. During *H. pylori* infection, the frequency of CD4+ T cells in the gastric lamina propria with a memory phenotype increases and polarizes to a Th1/Th17 phenotype, but these T cells are hyporesponsive to this bacterium [33]. Because this hyporesponsiveness contributes to chronic infection, attempts have been made by *H. pylori* to downregulate the T-cell response. *H. pylori* also manipulate T cell function by eliciting regulatory T cells (Tregs), frequently found in these patients [34]. Unusual Tregs activation by microbial antigens may provide a mechanism for preventing *H. pylori* from the immune response. Gamma-glutamyl transpeptidase (GGT) and VacA from the *H. pylori* molecule indirectly affect T lymphocyte activity and promote the differentiation of effector CD4+ T cells into Tregs. [35].

6. Current status and challenges of *H. pylori* vaccine candidate development

Since the early 1990s, vaccines based on various antigens, adjuvants, and routes of administration have been evaluated. The mucosal route of administration, especially the oral route, is the most suitable route for vaccination against *H. pylori* infection to induce an effective immune response at the site of infection. Until now, many vaccine candidates have been developed at the preclinical stage. In comparison, the most advanced candidate for the *H. pylori* vaccine is in phase 3 clinical trials (**Table 1**).

6.1 Oral recombinant *Helicobacter pylori* vaccine

This vaccine (UreB/LTB fusion vaccine) is a recombinant oral *H. pylori* vaccine using a urease B subunit (gene derived from *H. pylori* 9803) fused with heat-labile enterotoxin B subunit (gene derived from *E.coli* H44815) developed by Third Military Medical University and Chongqing KangWei Biotechnology in China. A randomized, double-blind, placebo-controlled phase III clinical trial was conducted in Ganyu County, Jiangsu Province, China. Vaccination was administered orally

Candidate vaccine	Country of the laboratory	Trial status	Prophylactic /therapeutic	References
Candidate vaccine: Oral Recombinant Helicobacter Pylori Vaccine	China	Phase III	Prophylactic	[36]
Imevax/IMX101	Germany	Phase I	Prophylactic	[37]
HelicoVax	USA	Preclinical	Therapeutic	[38]
Recombinant CTB-UreI-UreB (BIB)	China	Preclinical	Prophylactic	[39]
Recombinant *Vibrio cholerae* expressing *H. pylori* HpaA antigen	Sweden	Preclinical	Prophylactic	[40]
CTB-Lpp20	China	Preclinical	Prophylactic /Therapeutic	[41]
MCRI (Murdoch Children Research Institute /Gastric Cancer Vaccines)	Australia	Preclinical	Prophylactic	[42]
H. pylori Vaccine (NCT00736476)	Germany	I/II	Prophylactic	[43]

Table 1.
Summary of H. pylori *vaccine development status.*

in 3 doses on days 0, 14, and 28. In this study, oral administration of the *H. pylori* vaccine provided good protection against *H. pylori* infection in children aged 6–15 years up to 1 year after vaccination. Although the estimated point of protection for the vaccine later shows a slight decrease in efficacy, overall safety can last up to 3 years. All side effects are mild and improve within 24 hours. The most common reaction is vomiting, followed by fever and headache [36].

6.2 Imevax/IMX101

Imevax has completed a phase I clinical trial with IMX101. The vaccine consists of the *H. pylori* antigen-glutamyltranspeptidase (GGT), an outer membrane protein, and a mucosal adjuvant. The main reason for the failure of previous vaccines to provide complete protection is the immune evasion strategy possessed by *H. pylori* [35].

One of the most important is GGT, which appears to have relatively immunosuppressive solid activity. Therefore, this vaccine aims to target and neutralize these defense mechanisms, potentially enabling a more effective protective immune response against other antigenic components of the vaccine. The phase I clinical trial of Imevax IMX101 was conducted in a multi-center, randomized, double-blind, and adjuvant-controlled study conducted on adult volunteers aged 18 to 50 years to evaluate safety, tolerability, and efficacy. Volunteers consisted of negative people for *H. pylori* and healthy people infected with *H. pylori*. IMX101 vaccine is administered sublingually and intradermally [37].

6.3 HelicoVax

In the study by Steven F. Moss et al., they designed two DNA vaccines, namely HelicoVax A and HelicoVaxB, each containing a set of 25 different HLA class II epitopes. Steven F. Moss et al. used C57BL/6 mice. At six weeks, mice were infected with *H. pylori* strain SS1 in 0.1 ml PBS, 3 times in 1 week. DNA vaccines are administered intramuscularly and intranasally. The test results show that there is promising therapeutic efficacy for the development of an epitope-based mucosal vaccine against *H. pylori*.

6.4 Recombinant CTB-UreI-UreB (BIB)

Epitope vaccines are a promising option for protection against *H. pylori* infection. Research conducted by Jing Yang et al. developed a multi-epitope vaccine by linking the cholera toxin B (CTB) subunit, two antigenic fragments of *H. pylori* urease I subunit (UreI20–29, UreI98–107), and 4 *H. pylori* antigenic fragments. Urease B subunit, (UreB12–23,UreB229–251,UreB327–400,UreB515–561) produces recombinant CTB-UreI-UreB (BIB). This vaccine's protective effect against *H. pylori* infection was evaluated in BALB/c mice. Significant protection against *H. pylori* infection was achieved in BALB/c mice immunized with BIB, rIB plus rCTB, and rIB. Induction of substantial protection against *H. pylori* was mediated by specific serum IgA and mucosal sIgA antibodies and a mixed response of Th1/Th2/Th17 cells. This multi-epitope vaccine can be a promising vaccine candidate that helps control *H. pylori* infection [39].

6.5 Recombinant Vibrio cholerae expressing *H. pylori* HpaA antigen

The vaccine was designed by constructing and characterizing the faster-growing O1 *Vibrio cholerae* strain of *H. pylori* as a vector for the *H. pylori* antigen that might be used as a vaccine strain against *H. pylori*. Joshua Tobias et al. developed the technology of over-expressing enterotoxigenic *E. coli* (ETEC) colonization factor antigens (CFs), the main virulence factor of ETEC, in heterologous bacterial strains including *V.cholerae*. Due to the extracellular nature of *H. pylori*, the bacteria colonize the epithelial surface and coat the gastric mucosal lining and areas of gastric metaplasia in the duodenum. Joshua Tobias et al. constructed a *V.cholerae* strain that overexpressed HpaA, as a surface antigen and *H. pylori*-specific lipoprotein known to mediate *H. pylori* colonization in the rat stomach and be a protective antigen against *H. pylori* infection in animal models, singly or in animal models. Concurrently with different ETEC CFs can promote bacterial binding to the small intestinal mucosa. Specific strains were developed and characterized to the level of surface expression of HpaA, and the capacity to induce an immune response against *H. pylori* in mice after oral immunization [40].

6.6 Lp220 vaccine

Epitope vaccine is a potential vaccine as a prophylactic and therapeutic vaccine against *H. pylori* infection. Lpp20 is one of the main protective antigens that trigger immune responses after *H. pylori* invades the host and is considered an excellent vaccine candidate for the control of *H. pylori* infection. This epitope vaccine consists of a mucosal adjuvant cholera toxin B subunit (CTB) and three Lpp20 epitopes that have been identified (one B cell epitope and two CD4+ T cell epitopes) for efficacy in mice. An epitope vaccine consisting of CTB, one B cell, and 2 CD4+ T cell epitopes of Lpp20 was prepared and named CTB-Lpp20, which is expressed in *Escherichia coli* and used for immunization BALB/c mice *via* intragastric. The CTB-Lpp20 epitope vaccine has good immunogenicity and immunoreactivity and can produce high specific antibodies against Lpp20 and the cytokines IFN-c and IL-17. In addition, CTB-Lpp20 significantly decreased *H. pylori* colonization in mice. This protection correlates with IgG, IgA, and sIgA antibodies and Th1-type cytokines [41].

6.7 MCRI (Murdoch children research institute/gastric cvaccines)

MCRI (Murdoch Children Research Institute) developed a new vaccine strategy that prevents *H. pylori*-induced inflammation. The vaccine candidates are recombinant HtrA, a-55 kDa protein, and the only serine protease produced by *H. pylori*. HtrA is expressed and secreted on the bacterial surface, and is essential for the survival of *H. pylori*. MCRI investigators have shown that mice vaccinated with HtrA protected against *H. pylori*-induced inflammation compared with controls. *H. pylori* HtrA destroys the epithelial barrier by cleaving E-cadherin thereby opening junctions between gastric epithelial cells. The leaky epithelium allows a number of bacteria to enter the epithelial barrier to the tissue, interact with immune system cells, and cause gastritis. MCRI investigators found that serum from mice vaccinated with HtrA neutralized HtrA protease activity *in vitro* [42].

6.8 *H. pylori* vaccine (NCT00736476)

The vaccine consists of three recombinant *H. pylori* antigens vacuolating cytotoxin A (VacA), cytotoxin-associated gene A (CagA), and neutrophil-activating protein (NAP) that prevent infection in animal models and are well tolerated and highly immunogenic in adults receiving healthy. In this phase 1/2 randomized, single-center, unsupervised, placebo-controlled study, healthy nonpregnant adults aged 18–40 years who were confirmed negative for *H. pylori* infection were randomly assigned (3:4) to three intramuscular doses placebo or vaccine at 0, 1, and 2 months [43]. Previously, three recombinant antigen vaccines were tested relevant to *H pylori* virulence—CagA, VacA, and NAP—in phase 1 clinical study [44]. Compared with placebo, the vaccine did not provide additional protection against *H. pylori* infection after challenge with CagA-positive strains, despite an increased systemic humoral response to key *H. pylori* antigens. The vaccine induces high-titer IgG antibodies specific for CagA, VacA, and NAP and a robust antigen-specific T cell response, but this is not sufficient to eliminate *H. pylori* [43].

7. Conclusion

The best preclinical results are obtained from vaccines that often induce a T-cell-mediated immune response rather than humoral immunity. Th1 and Th17 responses in the stomach are more protective. The mechanisms of *H. pylori* persistence and the utilization of multiple mechanisms to overcome adaptive immunity are recognized as essential barriers to vaccination. Several vaccines were successfully tested in different experimental animal models, but translation into an efficacious human vaccine was unsuccessful. A better understanding of the immune response generated by natural *H. pylori* infection and the mechanism by which the bacteria survives is needed for the development of human vaccines. Future vaccines for the prevention of *H. pylori* infection should be conducted in children, where infection occurs naturally. Therefore, prophylactic vaccines may need to be given to children in the first few years of life to reach the maximum number of target groups when uninfected, but the health benefits will emerge five decades later.

Acknowledgements

We would like to thank the Faculty of Medicine and Health, Universitas Muhammadiyah Jakarta for support of this work.

Conflict of interest

The authors declare no conflict of interest.

Author details

Rike Syahniar*, Dayu Swasti Kharisma and Rayhana
University of Muhammadiyah Jakarta, South Jakarta, Indonesia

*Address all correspondence to: ri.syahniar@umj.ac.id

IntechOpen

References

[1] Syahniar R, Mardiastuti, Syam AF, Yasmon A. Phylogenetic analysis of Helicobacter pylori 16S rRNA gene in gastric biopsy from patients with dyspepsia. Advanced Science Letters. 2018;**24**(9):6789-6792

[2] Tsay F, Hsu P. H. pylori infection and extra-gastroduodenal diseases. Journal of Biomedical Science. 2018;**25**(65):1-8

[3] Bray F, Ferlay J, Soerjomataram I, Siegel RL, Torre LA, Jemal A. Global cancer statistics 2018: GLOBOCAN estimates of incidence and mortality worldwide for 36 cancers in 185 countries. CA: A Cancer Journal for Clinicians. 2018;**68**(6):394-424

[4] JKY H, Ying Lai W, Khoon Ng W, MMY S, Underwood FE, Tanyingoh D, et al. Global prevalence of Helicobacter pylori infection: Systematic review and meta-analysis. 2017 Aug;**153**(2):420-429 [cited 12 September 2021]; Available from:. DOI: 10.1053/j.gastro.2017.04.022

[5] Wang W, Jiang W, Zhu S, Sun X, Li P, Liu K, et al. Assessment of prevalence and risk factors of helicobacter pylori infection in an oilfield Community in Hebei, China. BMC Gastroenterology. 2019;**19**(1):4-11

[6] Syam AF, Miftahussurur M, Makmun D, Nusi IA, Zain LH, Zulkhairi, et al. Risk factors and prevalence of Helicobacter pylori in five largest Islands of Indonesia: A preliminary study. PLoS One. 2015 [cited 12 September 2021];**10**(11):e0140186 Available from: https://journals.plos.org/plosone/article?id=10.1371/journal.pone.0140186

[7] Syahniar R, Wahid MH, Syam AF, Yasmon A. Detecting the Helicobacter pylori 16S rRNA gene in dyspepsia patients using real-time PCR. Acta Medica Indonesiana. 2019;**51**(1): 34-41

[8] Khoder G, Sualeh Muhammad J, Mahmoud I, Soliman SSM, Burucoa C. Prevalence of helicobacter pylori and its associated factors among healthy asymptomatic residents in the united arab emirates. Pathogens. 2019 Apr 1;**8**(2):44

[9] Konno M, Yokota S, Suga T, Takahashi M, Sato K, Fujii N. Predominance of mother-to-child transmission of Helicobacter pylori infection detected by random amplified polymorphic DNA fingerprinting analysis in Japanese families. The Pediatric Infectious Disease Journal. 2008 [cited: 11 September 2021];**27**(11):999-1003 Available from: https://pubmed.ncbi.nlm.nih.gov/18845980/

[10] Wahid MH, Yuliantiningsih L, Syahniar R, Syam AF, Yasmon A, Krisnuhoni E, et al. Helicobacter pylori infection in siblings: A case report. In: Yunir E, editor. Medical Case Report. New York: Nova Science Publishers Inc; 2020. pp. 227-232

[11] Kipritci Z, Gurol Y, Celik G. Antibiotic resistance results of Helicobacter pylori in a University Hospital: Comparison of the hybridization test and real-time polymerase chain reaction. International Journal of Microbiology. 2020;**2020**:1-5

[12] Park JY, Shin TS, Kim JH, Yoon HJ, Kim BJ, Kim JG. The prevalence of multidrug resistance of helicobacter pylori and its impact on eradication in Korea from 2017 to 2019: A single-center study. Antibiotics. 2020;**9**(10):1-11

[13] Wang D, Guo Q, Yuan Y, Gong Y. The antibiotic resistance of Helicobacter pylori to five antibiotics and influencing factors in an area of China with a high risk of gastric cancer. BMC Microbiology. 2019[cited: 12 September 2021];**19**(1):1-10 Available from: https://

bmcmicrobiol.biomedcentral.com/
articles/10.1186/s12866-019-1517-4

[14] Schubert JP, Gehlert J, Rayner CK, Roberts-Thomson IC, Costello S, Mangoni AA, et al. Antibiotic resistance of Helicobacter pylori in Australia and New Zealand: A systematic review and meta-analysis. Journal of Gastroenterology and Hepatology [cited 12 September 2021]; 2021;**36**(6):1450-1456 Available from: https://onlinelibrary.wiley.com/doi/full/10.1111/jgh.15352

[15] Li J, Deng J, Wang Z, Li H, Wan C. Antibiotic resistance of Helicobacter pylori strains isolated from pediatric patients in Southwest China. Frontiers in Microbiology. 2021;**0**:3647

[16] JG K, AH v V, EJ K. Pathogenesis of Helicobacter pylori infection. Clinical Microbiology Reviews. 2006[cited 12 September 2021];**19**(3):449-490 Available from: https://pubmed.ncbi.nlm.nih.gov/16847081/

[17] Ierardi E, Losurdo G, Mileti A, Paolillo R, Giorgio F, Principi M, et al. The puzzle of coccoid forms of Helicobacter pylori: Beyond basic science. Antibiotics. 2020 [cited 12 September 2021];**9**(6):1-14 Available from: /pmc/articles/PMC7345126/

[18] Phe. UK Standards for Microbiology Investigations-standards-for-microbiology-investigations-smi-quality-and-consistency-in-clinical-laboratories PHE Publications gateway number: 2015075 UK Standards for Microbiology Investigations are produced in association with. 2015

[19] Kao CY, Sheu BS, Wu JJ. Helicobacter pylori infection: An overview of bacterial virulence factors and pathogenesis. Biomedical Journal. 2016 [cited 12 September 2021];**39**(1):14-23 Available from: https://pubmed.ncbi.nlm.nih.gov/27105595/

[20] Bauer B, Meyer TF. The human gastric pathogen Helicobacter pylori and its association with gastric cancer and ulcer disease. Ulcers. 2011;**18**(2011):1-23

[21] Kim N. Prevalence and transmission routes of H. pylori. *Helicobacter pylori*. 2016:3-19

[22] Oleastro M, Ménard A. The role of Helicobacter pylori outer membrane proteins in adherence and pathogenesis. Biology (Basel). 2013 [cited 13 September 2021];**2**(3):1110 Available from: /pmc/articles/PMC3960876/

[23] Shiota S, Suzuki R, Yamaoka Y. The significance of virulence factors in Helicobacter pylori. J Dig Dis. 2013 [cited 13 September 2021];**14**(7):341-349 Available from: https://pubmed.ncbi.nlm.nih.gov/23452293/

[24] Roesler BM, Rabelo-Gonçalves EMA, Zeitune JMR. Virulence Factors of Helicobacter pylori: A Review. Clinical Medicine Insights: Gastroenterology. 2014 [cited 13 September 2021];**7**(7):9 Available from: /pmc/articles/PMC4019226/

[25] Kalali B, Mejías-Luque R, Javaheri A, Gerhard M. H. pylori virulence factors: influence on immune system and pathology. Mediators Inflamm. 2014;2014:426309. doi: 10.1155/2014/426309. Epub 2014 Jan 21. PMID: 24587595; PMCID: PMC3918698

[26] Portal-Celhay C, Perez-Perez GI. Immune responses to Helicobacter pylori colonization: mechanisms and clinical outcomes. Clinical Science (London). 2006 [cited 13 September 2021];**110**(3):305-314 Available from: https://pubmed.ncbi.nlm.nih.gov/16464172/

[27] Phan Trung N. International PhD School in Biomolecular and Biotechnological Sciences Cycle XXVII Subject: Clinical and Molecular Microbiology. Polymorphisms of cagA and vacA Genes in *Helicobacter pylori*

Isolated from Gastroduodenal Diseases Patients in Central Vietnam Originality Statement Dissertation: INTERNATIONAL PhD SCHOOL IN Biomolecular and Biotechnological Sciences Cycle XXVII. Vietnam

[28] Suzuki H, Moayyedi P. Helicobacter pylori infection in functional dyspepsia. Nature Reviews Gastroenterology & Hepatology. 2013 [cited 13 September 2021];**10**(3):168-174 Available from: https://pubmed.ncbi.nlm.nih. gov/23358394/

[29] Kim Y-J, Chung WC, Kim BW, Kim SS, Il KJ, Kim NJ, et al. Is *Helicobacter pylori* associated functional dyspepsia correlated with dysbiosis? Journal of Neurogastroenterology and Motility. 2017 [cited 13 September 2021];**23**(4):504-516 Available from: https://www.jnmjournal.org/journal/view.html?doi=10.5056/jnm17066

[30] Chmiela M, Karwowska Z, Gonciarz W, Allushi B, Stączek P. Host pathogen interactions in *Helicobacter pylori* related gastric cancer. http://www.wjgnet.com/ 2017 [cited 13 September 2021];**23**(9):1521-1540. Available from: https://www.wjgnet.com/1007-9327/full/v23/i9/1521.htm

[31] Nouri A. Helicobacter pylori evasion strategies of the host innate and adaptive immune responses to survive and develop gastrointestinal diseases | Elsevier Enhanced Reader. Microbiological Research. 2019 [cited 13 September 2021];**218**:49-57 Available from: https://reader.elsevier.com/reader/sd/pii/S09445 01318308358?token=4B14C32FD720DDC 898F6AA5065BAD415012D776721C57A5 CE5A4414CB111011F16EA478013FEEA5 7B8759EC1CEC37B61&originRegion=eu-west-1&originCreation=20210913012435

[32] Stead CM, Beasley A, Cotter RJ, Trent MS. Deciphering the unusual acylation pattern of Helicobacter pylori lipid A. Journal of Bacteriology. 2008

[cited 13 September 2021];**190**(21):7012 Available from: /pmc/articles/PMC2580709/

[33] Lina TT, Alzahrani S, House J, Yamaoka Y, Sharpe AH, Rampy BA, Pinchuk IV, Reyes VE. Helicobacter pylori cag pathogenicity island's role in B7-H1 induction and immune evasion. PLoS One. 2015; [cited 14 Sepetmber 2021];**10**(3) Available from: https://pubmed.ncbi.nlm.nih.gov/25807464/

[34] Bagheri N, Shirzad H, Elahi S, Azadegan-Dehkordi F, Rahimian G, Shafigh M, et al. Downregulated regulatory T cell function is associated with increased Peptic Ulcer in *Helicobacter pylori*-infection. Microbial Pathogenesis. 2017;**110**:165-175

[35] Oertli M, Noben M, Engler DB, Semper RP, Reuter S, Maxeiner J, et al. Helicobacter pylori γ-glutamyl transpeptidase and vacuolating cytotoxin promote gastric persistence and immune tolerance. Proceedings of the National Academy of Sciences of the United States of America. 2013;**110**(8): 3047-3052

[36] Zeng M, Mao X-H, Li J-X, Tong W-D, Wang B, Zhang Y-J, et al. Articles efficacy, safety, and immunogenicity of an oral recombinant Helicobacter pylori vaccine in children in China: a randomised, double-blind, placebo-controlled, phase 3 trial. Lancet. 2015 Oct 10;**386**(10002):1457-64 [cited 14 Septermber 2021] . DOI: 10.1016/S0140-6736Available from:

[37] Phase 1a/b Study on Safety of IMX101 in H. Pylori-negative and H. Pylori-infected Healthy Volunteers - Full Text View - ClinicalTrials.gov. [cited 14 September 2021]. Available from: https://clinicaltrials.gov/ct2/show/NCT03270800

[38] Moss SF, Moise L, Lee DS, Kim W, Zhang S, Lee J, et al. HelicoVax: Epitope-based therapeutic H. pylori

vaccination in a mouse model. Vaccine. 2011;**29**(11):2085-2091

[39] Wang B, Pan X, Wang H, Zhou Y, Zhu J, Yang J, et al. Immunological response of recombinant H. pylori multi-epitope vaccine with different vaccination strategies. International Journal of Clinical and Experimental Pathology. 2014 [cited 14 september 2021];**7**(10):6559-6566 Available from: https://europepmc.org/articles/PMC4230088

[40] Tobias J, Lebens M, Sun Nyunt W, Holmgren J, Svennerholm A-M. Surface expression of Helicobacter pylori HpaA adhesion antigen on Vibrio cholerae, enhanced by co-expressed enterotoxigenic Escherichia coli fimbrial antigens. Microbial Pathogenesis. 2017 [cited 15 September 2021];**105**:177-184 Available from: https://pubmed.ncbi.nlm.nih.gov/28215587/

[41] Li Y, Chen Z, Ye J, Ning L, Luo J, Zhang L, Jiang Y, Xi Y, Ning Y. Antibody production and Th1-biased response induced by an epitope vaccine composed of cholera toxin B unit and Helicobacter pylori Lpp20 epitopes. Helicobacter. 2016 [cited 15 September 2021];**21**(3):234-248 Available from: https://pubmed.ncbi.nlm.nih.gov/26332255/

[42] Sutton P, Boag JM. Status of vaccine research and development for Helicobacter pylori. Vaccine. 2019 Nov;**37**(50):7295-7299. DOI: 10.1016/j.vaccine.2018.01.001

[43] Malfertheiner P, Selgrad M, Wex T, Romi B, Borgogni E, Spensieri F, et al. Articles efficacy, immunogenicity, and safety of a parenteral vaccine against Helicobacter pylori in healthy volunteers challenged with a Cag-positive strain: A randomised, placebo-controlled phase 1/2 study. The Lancet Gastroenterology & Hepatology. 2018 Oct;**3**(10):698-707 [cited 15 September 2021] Available from: www.thelancet.com/gastrohepPublishedonline

[44] Malfertheiner P, Schultze V, Rosenkranz B, Kaufmann SH, Ulrichs T, Novicki D, Norelli F, Contorni M, Peppoloni S, Berti D, Tornese D, Ganju J, Palla E, Rappuoli R, Scharschmidt BF, Del Giudice G. Safety and immunogenicity of an intramuscular Helicobacter pylori vaccine in noninfected volunteers: A phase I study. Gastroenterology. 2008 [cited 15 september 2021];**135**(3):787-795 Available from: https://pubmed.ncbi.nlm.nih.gov/18619971/

Chapter 6

Eg95: A Vaccine against Cystic Echinococcosis

Arun K. De, Tamilvanan Sujatha, Jai Sunder,

Prokasananda Bala, Ponraj Perumal,

Debasis Bhattacharya and Eaknath Bhanudasrao Chakurkar

Abstract

Hydatidosis or cystic echinococcosis (CE) is caused by the larval stage of the tapeworm Echinococcus granulosus. This parasite is cosmopolitan in distribution and causes significant economic losses to the meat industry, mainly due to condemnation of edible offal. Echinococcosis treatment in human is very expensive as it requires extensive surgery or prolonged chemotherapy or use of both. In Asia and Africa, the vulnerable population of developing the disease is around 50 million. Office International des Epizooties (OIE) has recognized CE as a multi species disease. The parasite has acquired the capability to survive long time within the host due to a specific mechanism to evade the host immune system. A specific class of proteins known as secreted and membrane bound (S/M) proteins play key roles in the evasion mechanism. A total of 12 S/M proteins have been reported as immunodiagnostic and immunoprophylactic agents. Of these, Eg95 is a candidate antigen used for immunization of animals. Literature suggests that, Eg95 is a multigene family (Eg95-1 to Eg95-7) and exists in seven different isoforms. This chapter will describe minutely efficacy of Eg95 as a vaccine candidate based on animal trial and potentiality of other S/M proteins as immunodiagnostic antigen and immune evasion.

Keywords: *Echinococcus granulosus*, cystic echinococcosis, secreted and membrane-bound (S/M) proteins, vaccine

1. Introduction

Echinococcus granulosus is the etiological agent of cystic hydatid disease (CHD) *alias* cystic echinococcosis (CE). CE is a classic example of cyclozoonosis since for completion of its life cycle, the parasite exploits two vertebrate hosts. The disease is an important cause of human morbidity and mortality specifically among transhumance pastoralists [1] and of worldwide distribution. In case of human, the disease poses a significant burden on health system due to the high cost of treatment including surgery and chemotherapy. Moreover, the disease has negative impact on productive and reproductive performances of farm animals in terms of reduction in production of milk, meat, and wool [2]. The global report suggests that human CE infection ranges from less than 1 per 100,000 to more than 200 per 100,000 in rural population. Prevalence of infection depends on association of man and dog. For zoonotic importance of the parasite and losses in livestock sector due to this infection, *Echinococcus* infection has been listed in the OIE Terrestrial Animal Health Code and is a notifiable disease for reporting by member countries and territories as per OIE code. In a most cited literature [2], disability-adjusted life years (DALYs) have been estimated as US $193,529,740 (95% CI, $171,567,331–$217,773,513). An annual livestock production loss of US $141,605,195 (95% CI, $101,011,553–$183,422,465) and possibly up to US $2,190,132,464 has been estimated as well. This has initiated the need to formulate control strategies. Guidelines of control measures include prevention of access of dogs to livestock carcasses, treatment of dogs with suitable anthelmintic, thorough meat inspection and disposal as well destruction of infected viscera and vaccination with Eg95 vaccine (http://www.oie.int/en/disease/echinococcosis).

2. Secreted and membrane-bound (S/M) proteins for sustenance within host: a weapon of parasite against host environment

We felt it judicious that, before we discuss about an Eg95 vaccine and secreted and membrane-bound (S/M) proteins of *E. granulosus*, let us brief S/M proteins in general since these biomolecules is the main functionaries for spicing up the immune system. As a general rule, helminth infection is a chronic infection because these harmful creatures survive in the host by its unique feeding strategies and evasion/muting of the host immune system. The weapon they use for winning the battle against the host immune system is S/M proteins (S/M) [3]. These unique biomolecules (S/M proteins) are associated with multitudinous activities such as

S. No.	Name of the parasite	Name of S/M protein	Activity
1.	*Schistosoma mansoni*	Chemokine binding Protein	Neutralization of chemokine activity (CXCL8, CCL3, CX3CL1, CCL2, CCL5); inhibits neutrophil migration [5] et al.
2.	*Fasciola hepatica*	Helminth defense molecule-1	Molecular mimicry of antimicrobial peptides, binds to LPS and reduces its activity; prevents acidification of the endolysosomal compartments and antigen processing; prevents NLRP3 inflammasome activation [6, 7].
		Fatty acid binding protein	Suppresses LPS-induced activation via binding and blocking of CD14; induction of alternatively activated macrophages [8]
		TGF-like molecule	Ligates mammalian TGF-b receptor (albeit with a lower affinity) and induces IL-10 and Arginase in macrophages [9]
3.	*Brugia malayi*	Asparaginyl-tRNA synthetase	Structural homology to IL-8, binds IL-8 receptors CXCR1 and CXCR2; chemotactic for neutrophils and eosinophils; induced regulatory responses and IL-10 in a T cell transfer model of colitis [10]
		TGF-b homolog-2	Ligates mammalian TGF-b receptor and suppresses T cell responses [11]
		Abundant larval transcript	Inhibitor of IFN-¥ signaling [11]

S. No.	Name of the parasite	Name of S/M protein	Activity
4.	*Necator americanus*	Metalloproteinases	Causes proteolysis of eotaxin, but not of IL-8 or eotaxin-2 [12]
		Ancylostoma secreted protein-2	Binding to CD79A on B cells, downregulation of lyn,PI3K, and BCR signaling [13]
5.	*Necator brasiliensis*	Acetylcholinesterase	Degrades acetylcholine, reduces neural signaling; induces proinflammatory cytokines with diminished type 2 cytokines in transgenic AChE-expressing trypanosome infection [14]
6.	*Echinococcus multilocularis*	T cell immunomodulatory protein	Induces release of IFN-g from CD4+ T cells in vitro [15]

Table 1.
Secreted and membrane-bound (S/M) protein of some important helminths of man.

penetration and establishment in the host; modulation of the host immune system and uptake of metabolites from the host [4]. Due to continuous exposure of S/M proteins with the host immune system, some of the candidate biomolecules exhibit immunodiagnostic or immunoprophylactic activities. Here we provide some of the examples of S/M proteins of helminths involved in survival strategies within the challenging host environment (**Table 1**).

3. Introducing S/M proteins of *E. granulosus*

This is better to understand about S/M proteins of *E. granulosus* because S/M proteins are used for control of the disease. We will not make the list long for the convenience of the readers and will zoom down to only four of them. Out of four, two are diagnostic antigens (Antigen B and Antigen 5) and two (Eg95 and 14-3-3 protein) are of immunoprophylactic value. Under this subheading, "Introducing S/M proteins of E. granulosus," we will brief on three S/M proteins except for Eg95, which we will elaborate later in this chapter.

3.1 Antigen B

Antigen B (AgB) is an oligomeric thermostable lipoprotein. The antigen was first described by Oriolet al. [16]. The protein was separated from the hydatid fluid by size-exclusion chromatography as a 160 kDa protein. This protein is abundantly present in *E. granulosus* hydatid fluid. AgB has already been characterized as an immunomodulatory protein, capable of inducing a permissible immune response to the parasite development. This protein is an oligomeric lipoprotein composed of 8 kDa related subunits [17]. Molecular studies revealed that AgB is encoded by a gene family with five major gene clusters, namely AgB1 [18], AgB2 [19], AgB3 [20], AgB4 [21], and AgB5 [22]. Bhattacharya et al. [23] carried out an exhaustive study on AgB families of Indian isolates of *E. granulosus*. AgB1 revealed homology to *Echinococcus canadensis* (G8) and *E. granulosus* sensustricto (G1/G2). AgB3 was homologous to *Echinococcus ortleppi* (G5) alias cattle strain. Predicted amino acid sequence of AgB4 was homologous to bovine isolates identified earlier.

3.2 Antigen 5

Antigen 5 (Ag5) is a major antigen of *E. granulosus*. This has been identified to have immunodiagnostic value. This is also known as Capron's arc 5 antigen because this antigen formed an arc by immunoprecipitation reaction with the serum samples of patients suffering from the disease [24]. Progress in the molecular characterization of Ag5 has been limited. Ag5 is a thermo labile protein. By size-exclusion chromatography, this has been eluted as 60–70 kDa protein [25]. By discontinuous gel electrophoresis, this has been found that, in native form Ag5 has a major component (67 kDa) and a minor component (57 kDa). Under reducing condition, Ag5 dissociates in two major peptides of 38 kDa and 22 kDa [26, 27]. Ag5 is a major component of hydatid cyst fluid, which is suggestive of its role as a key molecule in the biology of *E. granulosus*. This antigen plays an important role for development and sustenance of parasite within intermediate host till the transmission of the parasite to the definitive host [25].

3.3 14-3-3 Protein

14-3-3 proteins are a group of molecules that are of different isoforms. These molecules are distributed in a broad range of cells in all eukaryotic organisms. These groups of molecules are highly conserved in nature and were first reported from brain tissue. In recent time, they were found to play crucial roles in eukaryotic cell cycling. 14-3-3 Proteins bind with specific ligands containing phosphorylated serine/threonine residues to form homo- and heterodimer complexes, and this process is regulated by phosphorylation. Several mechanisms of action of 14-3-3 proteins have been reported; such as induction of conformational change of target molecules, the physical occluding of specific features, the scaffolding, and the change of cellular localization. The 14-3-3 proteins are acidic protein with a relative molecular weight of 30 kDa. This group of proteins show 50% identity within and across species, small (30 kDa), acidic proteins that show about 50% amino acid identity both within and across species. In mammals, seven isoforms have been identified (b-beta, c-gamma, f-zeta, r-sigma, e-epsilon, g-eta, and s-tau). The 14-3-3f isoforms also termed as *E14t* have been identified in *E. multilocularis* (Gen Bank accession no. U63643) and *E. granulosus* (Gen Bank accession no. AF20790). 14-3-3 Proteins have been found in metacestode, oncospheres, and protoscoleces of *E. multilocularis*. In *E. granulosus* these proteins have been found in protoscoleces and have potential role in the biology of this parasite. In one of the studies from India by Pan et al. [28], this was found that, there was over expression of 14-3-3 protein (zeta isoform) in drug induced protoscoleces of *E. granulosus* comprised to control group. This particular finding indicated that this protein may be used as biomarker in drug-induced protoscoleces.

4. Eg95: a brief introduction

Concept to develop Eg95 (16.6 kDa protein) was initiated on the basis of identification of individual oncosphere components that stimulate host-protective immune responses in sheep. Marathon effort was made on this aspect by Heath and Lawrence [29] on identification and characterization of host protective antigen hither-to its testing in vaccine trial. For raising the hyperimmune sera, the group of workers used whole *E. granulosus* oncospheres; non-denatured oncosphere extract treated by freezing, thawing and sonication; extract of immature oncospheres; denatured extract of oncospheres. By using an Geenzyme linked immunoblot assay

(EITB), a doublet immunodominant peptide of 23 kDa and 25 kDa was identified. The fraction that contained the 23 and 25 kDa molecules was able to stimulate protection in sheep. These studies suggested that one or both of the 23 and 25 kDa somatic oncospheral antigens of *E. granulosus* were host-protective components even after denaturation. This was the first indenture of native Eg95 vaccine preparation.

4.1 Isoforms of Eg95 antigen

A protein isoform is known as protein variant. They rise from a single gene or a gene family. Protein isoforms are formed due to alternative splicings or variable use of promoter or sometimes may be due to post-transcriptional modification of a single gene [30, 31]. Agene family of Eg95 for common sheep strain (G1) of *E. granulosus* has been described by Chow et al. [32]. They is Eg95-1 (Gen Bank ID: AF134378), Eg95-2 (AF 199351-52), Eg 95-3 (AF199353), Eg 95-4 (AF199349), Eg 95-5 (AF 199350), Eg 95-6 (AF199347), and Eg 95-7 (AF199348). Out of seven members of Eg95 family, Eg95-7 is pseudo gene.

Based on phylogenetic analysis (**Figure 1**), this was evidenced that Eg95 gene family is having two clusters (Eg95 1-4 and Eg 95 5-6). From India, an elaborative study was done to know the genetic diversity of Eg95 [33]. A total of 24 isolates collected from cattle, buffalo, sheep, goat, human, and dog were analyzed. Genotypic characterization of these isolates revealed that all isolates belonged to G1 genotype except one buffalo isolate, which was characterized as cattle strain (G5). Phylogenetically, the Eg95 coding gene characterized from Indian isolates of *E. granulosus* belongs to the Eg95-1/Eg95-2/Eg95-3/Eg95-4 cluster.

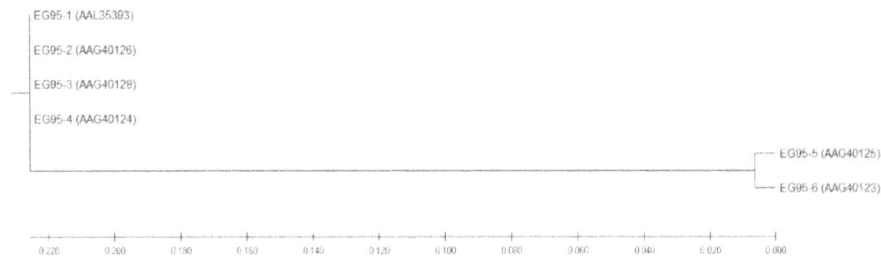

Figure 1.
Phylogenetic analysis of Eg95 family based on predicted amino acid sequence (Eg 95-7 could not be included since it is a pseudo gene).

4.2 Eg95 as vaccine

As a vaccine Eg95 is very effective to control *E. granulosus*. This is known that, after vaccination, there is antibody mediated and complement-mediated lysis of invading oncospheres. Initially let us mention a trial on vaccine, which was conducted in Rio Negro, Argentina. In the trial, lambs were vaccinated with Eg95 prepared by University of Melbourne, Australia. Primary immunization was done at 30 days of age and booster dose was applied at the age of 60 days. Final and penultimate dose was provided to the sheep at the age of 1–1.5 years. Immunological evaluation of vaccinated animals confirmed the presence of IgG antibody response, which persisted for a period of 5 years [34, 35]. Like sheep, Eg95 vaccine has been tested in cattle and goat. In cattle, after successful immunization, immunity persisted up to 12 months. This vaccine has been found safe and effective in pregnant sheep and cattle as well as in young small ruminants. A further elaborative study indicated that, after immunization with Eg95 vaccine in pregnant animals, there is passive transfer of maternal antibody response, which persisted for 3 months in lambs and 5 months in calves [36]. In a very recent note, immunoinformatics analysis and molecular docking tool have been employed to screen the antigen epitopes of *E. granulosus* with a novel purpose to design multi-epitope vaccine comprising of T and B cell epitopes. The multi-epitope vaccine was able to activate B lymphocytes to produce specific antibodies, which were predicted to confer protection in human being against the metacestode infection. This was further predicted that this multi-epitope vaccine was able to activate T lymphocytes and capable of immunological clearance. Further, four CD8[+] T cell epitopes and four B cell epitopes of *E. granulosus* tegument antigen were also predicted. Ultimately multi-epitope vaccine was predicted with the addition of linker protein [37].

Author details

Arun K. De, Tamilvanan Sujatha, Jai Sunder, Prokasananda Bala, Ponraj Perumal, Debasis Bhattacharya* and Eaknath Bhanudasrao Chakurkar
Animal Science Division, ICAR-Central Island Agricultural Research Institute, Port Blair, Andaman and Nicobar Islands, India

*Address all correspondence to: debasis63@rediffmail.com

IntechOpen

References

[1] Schantz PM, Chai J, Craig PS, et al. Epidemiology and control of hydatid disease. In: Thompson RCA, Lymbery AJ, editors. Echinococcus and Hydatid Disease. Oxon: CAB International; 1995. pp. 231-233

[2] Budke CM, Deplazes P, Torgerson PR. Global socioeconomic impact of cystic echinococcosis. Emerging Infectious Diseases. 2006;**12**(2):296-303

[3] Maizels RM, Smits HH, McSorley HJ. Modulation of host immunity by Helminths: The expanding repertoire of parasite effector molecules. Immunity. 2018;**49**(5):801-818

[4] Rosenzvit MC, Camicia F, Kamenetzky L, Muzulin PM, Gutierrez AM. Identification and intra-specific variability analysis of secreted and membrane-bound proteins from Echinococcus granulosus. Parasitology International. 2006;**55**(Suppl):S63-S67

[5] Smith P, Fallon RE, Mangan NE, Walsh CM, Saraiva M, Sayers JR, et al. Schistosoma mansoni secretes a chemokine binding protein with antiinflammatory activity. Journal of Experimental Medicine. 2005;**202**:1319-1325

[6] Robinson MW, Donnelly S, Hutchinson AT, To J, Taylor NL, Norton RS, et al. A family of helminth molecules that modulate innate cell responses via molecular mimicry of host antimicrobial peptides. PLoS Pathogens. 2011;**7**:e1002042

[7] Robinson MW, Alvarado R, To J, Hutchinson AT, Dowdell SN, Lund M, et al. A helminth cathelicidin-like protein suppresses antigen processing and presentation in macrophages via inhibition of lysosomalvATPase. FASEB Journal. 2012;**26**:4614-4627

[8] Martin I, Cabán-Hernández K, Figueroa-Santiago O, Espino AM. Fasciola hepatica fatty acid binding protein inhibits TLR4 activation and suppresses the inflammatory cytokines induced by lipopolysaccharide in vitro and in vivo. Journal of Immunology. 2015;**194**:3924-3936

[9] Sulaiman AA, Zolnierczyk K, Japa O, Owen JP, Maddison BC, Emes RD, et al. A trematode parasite derived growth factor binds and exerts influences on host immune functions via host cytokine receptor complexes. PLoS Pathogens. 2016;**12**:e1005991

[10] Kron MA, Metwali A, Vodanovic-Jankovic S, Elliott D. Nematode asparaginyl-tRNAsynthetase resolves intestinal inflammation in mice with T-cell transfer colitis. Clinical and Vaccine Immunology. 2013;**20**:276-281

[11] Gomez-Escobar N, Gregory WF, Maizels RM. Identification of tgh-2, a filarial nematode homolog of Caenorhabditiselegans daf-7 and human transforming growth factor b, expressed in microfilarial and adult stages of Brugiamalayi. Infection and Immunity. 2000;**68**:6402-6410

[12] Culley FJ, Brown A, Conroy DM, Sabroe I, Pritchard DI, Williams TJ. Eotaxin is specifically cleaved by hookworm metalloproteases preventing its action in vitro and in vivo. Journal of Immunology. 2000;**165**:6447-6453

[13] Tribolet L, Cantacessi C, Pickering DA, Navarro S, Doolan DL, Trieu A, et al. Probing of a human proteome microarray with a recombinant pathogen protein reveals a novel mechanism by which hookworms suppress B-cell receptor signalling. Journal of Infectious Diseases. 2015;**211**:416-425

[14] Vaux R, Schnoeller C, Berkachy R, Roberts LB, Hagen J, Gounaris K, et al.

Modulation of the immune response by nematode secreted acetyl cholinesterase revealed by heterologous expression in *Trypanosoma musculi*. PLoS Pathogens. 2016;**12**:e1005998

[15] Nono JK, Lutz MB, Brehm K. EmTIP, a T-Cell immunomodulatory protein secreted by the tapeworm Echinococcus multilocularis is important for early metacestode development. PLoS Neglected Tropical Diseases. 2014;**8**:e2632

[16] Oriol R, Williams JF, Pérez Esandi MV, Oriol C. Purification of lipoprotein antigens of Echinococcus granulosus from sheep hydatid fluid. American Journal of Tropical Medicine Hygiene. 1971;**20**(4):569-574

[17] Frosch P, Hartmann M, Mühlschlegel F, Frosch M. Sequence heterogeneity of the echinococcal antigen B. Molecular and Biochemical Parasitology. 1994;**64**:171-175

[18] Shepherd JC, Aitken A, McManus DP. A protein secreted in vivo by Echinococcus granulosus inhibits elastase activity and neutrophil chemotaxis. Molecular and Biochemical Parasitology. 1991;**44**:81-90

[19] Fernandez V, Ferreira HB, Ferna´ndez C, Zaha A, Nieto A. Molecular characterisation of a novel8-kDa subunit of Echinococcus granulosus antigen B. Molecular and Biochemical Parasitology. 1996;77:247-250

[20] Chemale G, Haag KL, Ferreira HB, Zaha A. Echinococcus granulosus antigen B is encoded by a gene family. Molecular and Biochemical Parasitology. 2001;**16**:233-237

[21] Isnd AC, Zaha A, Ayala FZ, Haag KL. The Echinococcus granulosusantigen B shows a high degree of genetic variability. Experimental Parasitology. 2004; **108**:76-80

[22] Haag KL, Arau´jo AM, Gottstein B, Siles-Lucas M, Thompson RC Zaha A. Breeding systems in Echinococcus granulosus (Cestoda; Taeniidae): Selfingor outcrossing? Parasitology. 1999;**118**:63-71

[23] Bhattacharya D, Pan D, Das S, Bera AK, Bandyopadhyay S, Das SK. An evaluation of antigen B family of Echinococcus granulosus, its conformational propensity and elucidation of the agretope. Journal of Helminthology. 2009;**83**(3):219-224

[24] Capron A, Vernes A, Biguet J. Le diagnostic immuno-e!lectrophore!tique de l'hydatidose. Lyon: SIMEP; 1967. pp. 27-40

[25] Bout D, Fruit J, Capron A. Purification d'un antige'nespe!cifique du liquidehydatique. Annual Reviews of Immunology (Paris). 1974;**125**:775-788

[26] Di Felice G, Pini C, Afferni C, Vicari G. Purification and partial charcterization of the major antigen of Echinococcus granulosus (antigen 5) with monoclonal antibodies. Molecular and Biochemical Parasitology. 1986; **20**:133-142

[27] Lightowlers M, Liu D, Haralambous A, Rickard M. Subunit composition and specificity of the major cyst fluid antigens of Echinococcus granulosus. Molecular and Biochemical Parasitology. 1989;**37**:171-178

[28] Pan D, Das S, Bera AK, Bandyopadhyay S, Bandyopadhyay S, De S, Rana T, Das SK, Suryanaryana VV, Deb J, Bhattacharya D. Molecular and biochemical mining of heat-shock and 14-3-3 proteins in drug-induced protoscolices of Echinococcus granulosus and the detection of a candidate gene for anthelmintic resistance. Journal of Helminthology. 2011;**85**:196-203

[29] Heath DD, Lawrence SB. Antigenic polypeptides of Echinococcus

granulosus oncospheres and definition of protective molecules. Parasite Immunology. 1996;**18**:347-357

[30] Brett D, Pospisil H, Valcárcel J, Reich J, Bork P. Alternative splicing and genome complexity. Nature Genetics. 2002;**30**(1):29-30

[31] Schlüter H, Apweiler R, Holzhütter HG, Jungblut PR. Finding one's way in proteomics: a protein species nomenclature. Chemistry Central Journal. 2009;**3**:11

[32] Chow C, Gauci CG, Cowman AF, Lightowlers MW. A gene family expressing a host-protective antigen of Echinococcus granulosus. Molecular and Biochemical Parasitology. 2001;**118**:83-88

[33] Sreevatsava V, De S, Bandyopadhyay S, Chaudhury P, Bera AK, Muthiyan R, et al. Variability of the *EG95* antigen-coding gene of *Echinococcus granulosus* in animal and human origin: implications for vaccine development. Journal of Genetics. 2019;**98**(2):53

[34] Larrieu E, Gavidia CM, Lightowlers MW. Control of cystic echinococcosis: Background and prospects. Zoonoses and Public Health. 2019;**66**(8):889-899

[35] Larrieu E, Mujica G, Araya D, et al. Pilot field trial of the EG95 vaccine against ovine cystic echinococcosis in Rio Negro, Argentina: 8 years of work. Acta Tropica. 2019;**191**:1-7

[36] Gauci C, Heath D, Chow C, Lightowlers MW. Hydatid disease: Vaccinology and development of the EG95 recombinant vaccine. Expert Review of Vaccines. 2005;**4**(1):103-112

[37] Yu M, Zhu Y, Li Y, Chen Z, Sha T, Li Z, et al. Design of a novel multi-epitope vaccine against echinococcus granulosus in Immunoinformatics. Frontiers in Immunology. 2021;**12**:66-84